ELECTRON
STRUCTURE AND
HIGH-TEMPERATURE
CHEMISTRY OF
COORDINATION
COMPOUNDS

CHEM

ELECTRON
STRUCTURE AND HIGH-TEMPERATURE CHEMISTRY OF COORDINATION COMPOUNDS

Yu. Buslaev (Editor)

Nova Science Publishers, Inc.
Commack

Art Director: Maria Ester Hawrys
Assistant Director: Elenor Kallberg
Graphics: Denise Dieterich, Kerri Pfister,
        Erika Cassatti and Barbara Minerd
Manuscript Coordinator: Sharyn Schweidel
Book Production: Tammy Sauter, Benjamin Fung
Circulation: Irene Kwartiroff and Annette Hellinger

*Library of Congress Cataloging–in–Publication Data
available upon request*

ISBN 1-56072-186-1

© *1996 Nova Science Publishers, Inc.*
  6080 Jericho Turnpike, Suite 207
  Commack, New York 11725
  Tele. 516-499-3103 Fax 516-499-3146
  E Mail Novasci1@aol.com

Printed in the United States of America

CONTENTS

Theoretical Study of Isomerism in Simple Complexes of Metal Cations With Ambivalent Ligands
O.P. Charkin and N.M. Klimenko

1. Introduction.. 1

2. Computational Approximation and Quality of Nonempirical
   Calculations 4

3. Potential Energy Surfaces and Isomerism of M[ABC] Salt
   Molecules and Ion-Molecular Complexes With Di- and Triatomic
   Ligands AB and ABC.. 7

3.1. Complexes M[AB] with Diatomic AB Ligands Containing 10 Valent
   Electrons... 7

3.2. Comparison of PES of Rearrangements MAB->ABM (M=Li and Cu)
   and HAB->ABH.. 12

3.3. Role of the Charge of Cation and its Chemical Environment................ 17

3.4. Fragments LiAB of Diatomic Ligands with 12 Valent Electrons........... 20

3.5. Complexes LiAB with 14 Valent Electrons 22

3.6. Complexes M[ABC] of Triatomic Ligands ABC with 16 Valent
   Electrons... 24

3.6.1. Chalkogenides YAY'(Y,Y'=0 and S,A=B, Al, C and Si, etc.).............. 24

3.6.2. Cyanates and Thiocyanates... 27

3.6.3. Other Ligands ABC With 16 Valent Electrons................................. 28
3.7. Complexes of Triatomic ABC Ligands With 18 and 20
   Valent Electrons 29

3.7.1 Anions NO_2^-, NS_2^-, PO_2^-, and PS_2^-, Molecules O_3SO_2, S_2O, S_3
   and Their Analogs.. 29

4.1 Ions M_2AB^+ With Two Cations 31

4.2 Ions M_2ABC 34

4.3 Dimers $(MAB)_2$ 36

5. Summary 38

6. References 40

7. Tables 49

8. Figures 65

Ferrocene-Based Metallochelates
V.A. Kogan and A.N. Morozov

Introduction 75

1. Transition Metal Complexes With Ferrocene-Diketones 76

2. Metallochelates Based on Thisemicarbazones and Hydrazones
   of Acyl Ferrocene Derivatives 85

3. Other Hetermetallic Ferrocene-Containing Metallochelates 89

4. The Problem of Structural Modification of Ferrocene by the Method of
   Chelate Formation 91

References 99

Figures 105

Cluster Compounds of Boron in Coordination Chemistry
N.T. Kuznetsov and K.A. Solntsev

Introduction 107

1. General Characteristics of Cluster Boron Compounds 108
1.1. Formation of Polyhedral Boron Hydride Structures 109

1.2. Structure and Peculiar Features of Chemical Bonds 111

1.3. Reaction Capacity..115

1.4. "Superelectrondeficient" Structures......................................117
Comparison of Polyhedrons $B_{10}H_{11}^-$ in Tetragonal and Monoclinic
Structures and Transition Between Them..............................122

2. Complex Compounds of Transition Metals With Outer-Sphere
Polyhedral Anions...125

2.1. Complex Compounds of Palladium (II) and Platinum (II)...................126

2.2. Complex Compounds of Mn(II), Fe(II), Co(II), Ni(II), Cu(II) and Ln(III)
....128

2.3. Complex Compounds of Uranyl UO^{2+} and U(IV................................132

3. Complex Closo-Dodecaborate-Halides....................................134

3.1. Physical and Chemical Studies of Systems and Synthesis of Complexes
....135

3.2. Structure and Crystal-Chemistry Stability Criterion........................136

3.3. Physical and Chemical Properties and Their Relationships With
Structure..138

4. Polyhedral Anions B_nH_n as Ligands......................................140

4.1. Complex Compounds of Pt(II), Pd(II) and Cu(I)..............................141

4.2. Complex Compounds of Pb(II)...143

4.3. Complex Anion $CuB_{10}H_{10}$..144

Conclusion...146

References..146

High Temperature Coordination Chemistry
S.V.Volkov

I. Introduction...151
II. Chemistry of Coordination Compounds in Ionic Metals....................156

1. Complex-Cluster Model of Structure of Molten-Salt Systems With Complex

Formation..**158**

2. Discrete Complex Ion in Melts as a Quadrinuclear Heteronuclear
System..**163**
3. Quantum-Chemistry and Experimental Questions of Structure and
Radiative and Nonradiative Transitions of Coordination Compounds in
Molten Salts...**167**

4. Some Common and Specific Problems of Ion Coordination in Molten Salts
and Molecules in Molecular High-Boiling Solvents..............................**179**

III. Chemistry of Coordination Compounds in Anhydrous High-Boiling
Solvents...**183**

1. Problem of Simultaneous Exhibition of Properties of Solvent-Ligand-
Reagent...**183**

2. Formation of Heteronuclear, Dinuclear and Other Complexes...............**189**

3. Some Solvato-Metallurgical Aspects of Problem..................................**196**

IV. Chemistry of Coordination Compound in the Vapor (Gas) Phase........**197**

1. Quantum Chemistry, Spectroscopy and Structure of Volatile
B-Diketonate Metals and Vapor(Gas) Phase [38,41,44,47,49,50,54].........**200**

2. Problem and Criteria of Volatility of B-Diketonates of Metals............**210**

3. Gas-Transport, Homophase, Heterophase Processes With
Participation of High-Volatile B-Diketonate Complexes [38]................**218**

V. Coordination Compounds in Laser Chemistry [58-72]............................**221**

1. Laser Chemistry of High Coordination Compounds in Gas Phase..........**222**

2. Laser Chemistry of Coordination Compounds in Molten Salts................**224**

3. Laser Chemistry of Coordination Compounds in Polymerizing and
SolidPhases..**227**

References..**233**

Subject Index...**237**

THEORETICAL STUDY OF ISOMERISM IN SIMPLE COMPLEXES OF METAL CATIONS WITH AMBIVALENT LIGANDS

O.P. Charkin and N.M. Klimenko

1. INTRODUCTION

In the last decade, the interest of chemists has become more and more attracted to the phenomena of "bonding" isomerism in complex compounds, which is manifestated in the ability of ligands $L = ABCD ...$, containing several atoms with electronic pairs, to be coordinated to the central metal cation M^{q+} by means of different donor atoms A, B or C , etc with the production of different isomers.

"Bonding" isomerism was discovered first by Jorgensen in 1894 [1] for the ammonia complexes of Cobalt (III) $[Co(NH_3)_5(NO_2)]^{2+}$ and $[Co(NH_3)_5-(ONO)]^{2+}$, where the NO_2^- anion can be coordinated to the Co^{3+} by hydrogen atom, as a hydrogroup, or by oxygen, as a hydrogroup. Numerous x-ray, IR and NMR spectroscopic investigations demonstrate the bonding isomerism to be widely distributed and typical for the cyanide CN^- (coordination by C and N), cyanate CNO^- and thiocyanate CNS^- (by N and O or S), formamide NH_2CHO (by S and O groups, for complexes with carbamide, thiosemicarbazide and other ambivalent ligands (see, for example [2–10] and references herein).

For understanding, even qualitative, of general features of the bonding isomerism on a molecular level, the following questions are of interest.

1. What types of "bonding" isomers can exist and be detected experimentally for complexes with different cations M^{q+} and ambivalent ligands L? What are their geometric and electronic structures, vibrational frequencies, dipole moments and other properties?

2. What is the energetic and kinetic stability of alternative bonding isomers in relation to their intramolecular rearrangements each into another (their relative energies and potential barriers between them)? What is the dependance of these characteristics on nature of the ligand L and cation M^{q+}, and on the chemical environment of the latter one in its coordination sphere?

3. When can the isomerization go along the intramolecular pathway and when can it go under the dissociative mechanism?

4. What is the role of the environment (solution, crystal lattice, and other factors) in the problem of relative stability of the bonding isomers and mechanisms of the bonding isomerization?

Attempts to answer these and other related questions on the experimental ground will only prove to be harmful because of the shortage of experimental data being scattered and far to be complete, as well as by limited possibilities of every individual experimental method taken separately. It seems to be obvious for the adequate study of the bonding isomerism it is necessary to use a proper combination of several experimental methods simultaneously, both structural and energetic ones, but cases of this type are rare in literature. For example, x-ray diffraction data is compiled for the thio- and selenocyanate complexes of many transition metals, where the NCS⁻ and NCSe⁻ anions are coordinated either by N or by O atoms, but for every compound under study only a single structure is characterized. In [5—7] the IR-spectroscopic data in solids is complied, but the cases where structural data is measured for both (or several) alternative isomers is rare.

Possibilities of the spectrophotometrical and NMR spectroscopic investigations in solutions appear to be more open [8—10]. Using these methods one could detect and demonstrate existence of many alternative bonding isomers with the ambivalent ligands L, coordinated by different donor atoms, but while defining properly a type of coordination of M^{q+} and L, and sometimes—a type of symmetry of the isomer with the M—L bond, they are unable usually to find equilibrium interatomic distances and valent angles.

Even more rare are experimental data for relative energetic stability of the isomers, for activation barriers and mechanisms of the isomerization reactions. Thus, with the use of the differential scanning calorimetry, the enthalpy of the reaction

$$[Co(NH_3)_5(NO_2)]Cl_2 \text{ -----} \rightarrow [Co(NH_3)_5(ONO)]CL_2$$

was evaluated about 2 kcal/mol [11], and of the reaction

$$K_4[Ni(NO_2)_6] \cdot H_2) \text{ -----} \rightarrow K_4[Ni(NO_2)_4(ONO)_2] \cdot H_2O$$

about 3.5 kcal/mol per every dimerizing NO_2^- group [12]. There are evidences that the same isomerization reactions can undergo different mechanisms in the solution and in crystal states, correspondingly[a]), that the pH of the solution can play an important role, etc.

As a whole, experimental data available for the bonding isomers so far are too scarce and fragmental for generalizations on a molecular level. In situations of this kind, systematical nonempirical quantum-chemical calculations of potential energy surfaces (here and below, PES) can be very helpful, if they are performed in the vicinity of the alternative geometric configurations of the complexes, where the ligand L is coordinated to the central M^{q+} cation by its different donor atoms, as well as of intermediate structures in the vicinity of the potential barrier, separating these isomers. The calculations can give information (on a semi-quantitative or even a quantitative level) on equilibrium geometry parameters of the isomers, their vibrational frequencies, relative energies and height of potential barriers. The relative energies are important to judge the energetic stability of the isomers, while the barriers-for judging of their kinetic stability to the mutual transformation each into another by the intramolecular mechanism.

The quantum-chemical approach has its own disadvantages and limitations. First of all, almost all calculations available now are performed for isolated systems and do not include entropy members, while experimental data of coordination chemistry is related to the systems in solution and in crystals at the normal temperature (or in other conditions different from O° K). In other words, calculations and experimental measurements are performed for different conditions of the systems under study. In spite numerous examples of their quite reasonable correspondence, their results, strictly speaking, must not always coincide. For the adequate comparison of experimental and calculated data, the computational approaches cannot be limited by investigation of the isolated complexes or molecular clusters cut out of the crystal, but to take into account explicitly effects of solvation or of the crystal lattice. Usually, in the theoretical approaches, for the discussion of the environment, a potential energy surface of the isolated system (complex or cluster) is taken as a starting point, while the influence of the solvent is treated as a perturbation. Unfortunately, as far as we know, calculations of this level are absent for the bonding isomerization rearrangements in complex compounds.

Moreover, semiempirical calculations of coordination compounds (even in the isolated state) appear to be not enough reliable, while the more precise non-empirical calculations with detailed scanning of potential energy surfaces are expensive. An overwhelming majority of the nonempirical calculations of

---

[a]For example, with using the isotope method, authors [13] concluded the isomerization of the thiocyanate group in the $[Co(NH_3)_5NCS]^{2+}$ complexes under the intramolecular mechanism, while in the crystal salt $[Co(NH_3)_5NCS](NCS)_2$ by exchange of the intra- and outerspheric NCS anions.

PES of the isomerization is performed for the simple isolated fragments cation-ligand of the M—L or L'_nM—L type, where the M^{q+} cation either is "naked," or is surrounded by a few simple ligands of the type of H_2), NH_3, CO, etc. Surely, this limitation—to study the simple fragments—is no more than the first step, the next ones are problems of the mutual influence of ligands and the role of the chemical environment of the cation M^{q+}, etc.). This way promises a chance to find the most typical peculiarities of PES'es of the isomerization rearrangements for different cations M^{q+} and ligands L, to follow their similarities and differences in rows of related systems, and to understand on a model level the most important regularities of structure and relative stability of the bonding isomers and their dependence on nature of the cation and anion. Moreover, it is worth to underline that the same simple fragments of the type of CuCN, SrNCO, CaN_3, CuNO, CoN_2, LiPO or $BaCN^+$, LiN_2^+, $LiCO_2^+$ as isolated molecules or ion-molecular complexes are of large interest for the gas-phase chemistry: in the last decade, many of them have been detected and investigated in high-temperature vapors (e.g., above the melted salts), in molecular beams, in inert matrix isolation conditions.

This review has a goal to generalize the results of *ab initio* calculations of potential energy surfaces of the bonding isomerization rearrangements in the M—L fragments with various cations M^{q+} and ligands L.

2. COMPUTATIONAL APPROXIMATION AND QUALITY OF NONEMPIRICAL CALCULATIONS

Before the analysis of results it is necessary to briefly discuss some of the calculations and the dependence of the calculated data on a computational approximation.

Usually, a scanning of potential energy surface (PES) along the minimal energy pathway (MEP) of migration of the M cation around the L ligand is performed in a following way. An angle $\varphi_{(MxA)}$ with a vertex x in the middle of diatomic ligand AB (Figure 1) or an angle $\varphi_{(MBA)}$ with a vertex in a central atom B of triatomic ligand ABC (Figure 14) is taken as a reaction coordinate and varied within an interval from 0° to 360°. At every fixed value of φ, all other geometric parameters are optimized on the SCF level. Thus, a profile of the PES along the MEP can be found, and a character of its special points and their vicinities understood. Further, the geometry of the special points is reoptimized and their energies are precized by the single-point calculations using more complete and flexible basis sets and taking into account the electronic correlation. The optimization of geometry of the majority of molecules and ion-molecular complexes with cation M^{q+} of non-transition metals discussed below was performed in the SCF approximation with the $3-21G^*$ or $6-31G^*$ basis sets including polarizing 3d-functions on all atoms except the alkali cations Li^+, Na^+, etc. In the single-point calculations of energies, the $6-31G^*$ or Huzinaga–Dunning double-zeta basis sets +

polarization (DZHD + P [14,15] were used), while electronic correlation was taken into account for several (not all) typical molecules within the Möller–Plessett perturbation theory of the 3rd and 4th order (MP3 and MP4), configuration interaction (CI), self-consistent electronic pairs (SCEP) and other methods (see notes to Tables below). By experience of computations of relative (but more covalent) compounds, the approximation mentioned above describes experimental values of equilibrium internuclear distances within a few hundreds of angström, a few degrees for valent angles and a few (sometimes 5) kcal/mol for relative energies of alternative structures. Unfortunately, chances to compare computational results for the isolated salt molecules and ion-molecular complexes with polar cation-ligand bonds directly with experimental data are limited today because of the scarcity of latters. For the system where experimental data is available, the agreement between calculations and the experiment remains to lie approximately within the same limits (and even less for relative energies) as for the related more covalent species.

Calculations of compounds with cations M^{q+} of transition metals are more complicated and approximate. Here we must limit ourselves in order to consider isomerization which is mainly used for complexes of Cu^+ and Zn^{2+} with a closed d^{10} shell where the triplet states lie high enough in energy and the wave function of the ground singlet state can be described adequately by a single determinant. Scanning and optimization of special points of their PES (see Tables 2, 4B and 8B) were carried out within the SCF approximation using the double-zeta Roos–Veillard–Vinot (DZRVV) basis set for 3d atoms [16] + Roos–Siegbahn (DZRS) basis set for non-transition elements [17]. Energies of the special points of PES were precized using the double-zeta Wachter's (DZW) basis set for the transition metals [18] and Huzinaga–Dunning (DZHD) basis set for the non-transition elements [14–15]. Comparison of results of calculations carried out with various basis sets between each other (see [19] and references herein) and with experimental data demonstrates the errors of this approximation can be up to 0.05–0.10 Å for the internuclear distances and 5–10° for the valent angles. Results of calculations [19] within this approximation for molecules CaF_2, TiO_2, $CuCN$, $CuCO^+$ and CuH_2^+ are in a good agreement with later calculations [20–24] performed with more flexible basis sets and the electronic correlation taken into account. As for the complexes with the non-closed d^n-shell, their non-empirical calculations appear to be much more complicated because of the high density of the low-lying excited electronic states, problem to find the ground electronic state (it can be changed during the isomerization). It is necessary to use many-determinant reference wave functions here, but the calculations on this level have started just recently.

A special case represents a so-called migrational or deformational "non-rigid" molecules having a very flat potential energy surface within a wide interval of the angle φ. Their alternative isomers are close in energy and divided by low barriers which can be overcome at the normal or moderately-high temperatures [25–30]. The obtainment of the equilibrium values of their geometry parameters r_e, φ_e appears to be a particular delicate and difficult problem both for the computational chemistry and for the experimental structural methods. In the quantum-chemical calculations it is necessary to use a more complete and flexible basis sets when optimizing geometrty (for example, it was shown in [31–33], that the DZHD + P basis set with the single-zeta polarizing 3d-function is not sufficient to reproduce adequately a cyclic structure of the NaCN and KCN cyanide molecules, measured in [34,35] by the microwave spectroscopy method; the agreement becomes satisfactory only with using the double-zeta polarizing functions. Similar results were found for RbCN [36,37], while for the lithium cyanide the linear isocyanide LiNC structure was justified by both calculations [38] and microwave measurement [39] and sometimes took into account the electronic correlation. On the other hand, the relative energies of isomers and potential barriers in complexes with highly polar bonds cation—ligands usually depend relatively weakly on the electronic correlation, and the SCF-approximation, as a rule, is enough for the adequate description of the relative energetic characteristics, at least on the semi-quantitative level.

As for experimental measurements of structure of the nonrigid molecules, it is worth it to underline that usually experiment gives the "effective" values of geometric parameters r_g, φ_g which are averaged over all occupied vibration levels at high temperature conditions (e.g., in vapors above the melted salts); they take into account effectively all types of interactions with the matrix atoms under inert matrix isolation conditions, etc. For the structural-rigid molecules having well-defined minima which are separated by enough high barriers, the difference between the equilibrium r_e, φ_e and effective r_g, φ_g values of geometry parameters is small and usually can be neglected. Contrarily, for the structurally non-rigid systems this difference can be essential. Moreover, different experimental methods sometimes can give different results for the same non-rigid molecule. For the adequate comparison of calculated structural data with experiment, as well as data of different experimental methods between each other, it is necessary to recalculate the effective r_g, φ_g values, measured at certain conditions, to the equilibrium r_g, φ_g. As long as these recalculations are not done, the calculated and experimental data can differ from each other not only because they have errors, but also because of their calculations and experiments which characterize the same molecule being in different conditions.

Taking into account these circumstances, in the present review we do not put a goal to discuss quantitatively the absolute values of the structural and energetic characteristics of every molecule or ion-molecular complex, taken individually. We limit ourselves by the semi-quantitative approach and concentrate our attention mainly on the tendencies and regularities of the characteristics of the bonding isomers in series of related molecules and ions. We believe that the errors of calculations within the same computational approximation to be of systematical character in series of similar species.

When analyzing the tendencies of molecular properties, the systematical errors can be essentially compensated resulting in a more higher precision for the relative values as compared to the absolute values of the properties for every member of the series taken separately.

It is necessary to be cautious when discussing "delicate" calculated effects (e.g., shallow local minima and low barriers about 1–2 kcal/mol or less) which can sometimes be met on the flat sections of the potential energy surfaces of the non-rigid species. Precision of the computational approximations mentioned above can not be high enough to determine if these effects are definitely artifacts of the calculations or of the real physical nature. More precise calculations and experimental investigations are necessary here.

3. POTENTIAL ENERGY SURFACES AND ISOMERISM OF M[AB] AND M[ABC] SALT MOLECULES AND ION-MOLECULAR COMPLEXES WITH DI- AND TRIATOMIC LIGANDS AB AND ABC

We will start from the potential energy surfaces (PES) of the isolated salt molecules or ion-molecular complexes M[AB] and M[ABC] containing one cation and one diatomic AB or triatomic ABC ligand (anion or neutral molecules). Many of them were detected and studied mass-spectrometrically in the gaseous phase [40–51] or in the inert matrix isolation [52–67]. Their structure was investigated by the IR spectroscopy [52–67], electron diffraction [68–72], laser spectroscopy [73–75] and other methods. Apart from individual interests (as isolated molecules and ions), they can be regarded as fragments cation—ligand M—AB and M—ABC cut out from the coordination sphera of more complicated complex compounds, and taken as suitable simple models for study of the bonding isomerism without complications of the ligands mutual influence effects (the cation is "naked") and the environment (the fragment is isolated).

3.1. Complexes M[AB] with Diatomic AB ligands Containing 10 Valent Electrons

Molecules of salt MAB and ion-molecular complexes MAB^+ with diatomic ligands AB^- (anions or neutral molecules),[a] containing 10 valent electrons, were studied by many authors. The most complete collection of the

---

[a]Here and below we will designate A and B the atoms disposing in the left and, correspondingly, in the right halves of the Periodic Table. As a rule A is a more electropositive and B a more electronegative atom in the AB ligand. The Roman numeral (the upper index) designates the period of the Periodic Table corresponding to given ligand's atom A (or B).

potential energy surface, calculated in good methodical approximations is available for them. Typical profiles of PES along the minimal energy pathway (MEP) of migration of the M^{q+} cation and the AB ligand relative each other are plotted on Figure 1 (compounds of the Li^+ cation are investigated most thoroughly and in detail), and the alternative structures, corresponding to their special points, on Figure 2. Table 1 contains geometric parameters optimized within the SCF approximation using the 3-21G* and DZHD + P basis sets, and relative energies of special points of PES, precized in the SCF, MP3 or CISD approximations with the basic sets of the DZHD + D type [76–89].

The most thoroughly were studied the complexes and salts of the alkali and alkali-earth cations with light ligands $A^{II}B^{II}$ = CO, N_2, CN$^-$, NO$^-$, etc. containing atoms from the second period (first row). One can see from Figure 1 and Table 1 that the compounds of the light ligands CO, N_2, CN$^-$, BO with N_{val} = 10 have very flat PES, typical for the structural non-rigid systems [25–30]. The $A^{II}B^{II}$ anions and molecules are coordinated to the cation Li^+ as δ-ligands only either by electropositive A^{II} or by electronegative B^{II} atoms producing two linear isomers Li—A^{II}—B^{II} and A^{II}—B^{II}—Li correspondingly. The intermediate bent and cyclic structures with Li^+ located in the vicinity of the A^{II} atom or above the A^{II}—B^{II} bond correspond either to the top of the migration barrier or to the non-stationary points of PES. The carbonyl group is coordinated mainly by its carbon end, the cyanide group by the nitrogen atom (in accordance with the microwave spectroscopy data for CNLi [39]), but their excited isomers lie only 3–5 kcal/mol higher. For the BO$^-$ group, the difference is even less, and keep in mind the approximate character of the calculations, it seems reasonable to speak of the energetic quasi-degeneration of isomers LiBO and BOLi. The barriers dividing the linear isomers are not high, and one can assume the Li^+ cation and the $A^{II}B^{II}$ ligand to be able, not only to perform the large-amplitude vibrations, but to migrate around each other at moderately high temperatures. The migrational non-rigidity of the systems like CNLi was discussed in detail in [25–29,78]. The cation—ligand bond here is of polytopic character. When migration of the cation, different types of intramolecular interactions each transform into another continuously, their energetic contributions change in a balanced way and compensate each other rather well, although surely, but not completely. As a result, all the changes of their total energy lie within narrow limits along the all migration pathway. The disbalance effects increase rapidly when increasing the difference in electronegativity of the ligand's atoms A and B, and coordination of Li^+ to the more electropositive A and B atoms in the LiBF$^+$ and LiBeF species comes to be more (by 15–20 kcal/mol) advantageous as compared to the coordination of the electronegative F end. Systems of this kind remain to be "local" non-rigid to the large amplitude movements in the vicinity of the equilibrium structure, nevertheless the migration of the cation around all the space of the BF or BeF$^-$ system appears to be nonprofitable energetically.

It is worth it to underline the $LiCO^+$ and LiN_2^+ ion-molecular complexes are more rigid when compared to the isoelectronic neutral $LiBO$ and $LiCN$ salt molecules; the migration barriers h_{migr} for the ions are in 1.5–2.0 times higher than for the molecules [87–89].

The migration non-rigidity of the carbonyl and cyanide complexes is of interest to the bonding isomerism in two aspects. Firstly, closeness in energy of the $Li—A^{II}B^{II}$ and $A^{II}B^{II}—Li$ isomers is an evidence in favor of the ability of these ligands to be coordinated by both donor atoms A and B. Secondly, small potential barriers between the isomers do not exclude the possibility of the isomerization to undergo intramolecular rearrangement (under the monomolecular mechanism).

The e is a question, how will it change PES, relative stability and structure of the isomers and height of potential barriers when replacing the light A^{II} and B^I atoms from the second period, one by one in a consecutive order of both simultaneously, by their more heavier analogs along the subgroup from the third period, e.g., $A^{II} \to A^{III}$ and $B^{II} \to B^{III}$.

As it can be seen from Figure 1 and Table 1, that the balance and compensation of the energetic contributions of different intramolecular interactions, which are characteristic of the compounds of the light $A^{II}B^{II}$ ligands and very important for their migrational non-rigidity, are violated when replacing one of the ligand's atoms and passing from the $LiA^{II}B^{II}$ to $LiA^{III}B^{II}$ or $LiA^{II}B^{III}$. The $LiA^{III}B^{II}$ and $A^{II}B^{III}Li$ structures with Li^+ coordinated to the "heavy" A^{III} or B^{III} atoms are strongly destabilized, while the $LiA^{II}B^{III}$ and $A^{III}B^{II}Li$ structures with Li^+ coordinated to the "light" A^{II} or B^{II} atoms are stabilized. As a result, as compared to the $LiA^{II}B^{II}$ $A^{II}B^{II}Li$ rearrangements, those sections of PES, corresponding to movement of Li^+ in vicinity of the heavier atom, are sharply shifted up on the energy scale, while the sections with Li^+ in the vicinity of the light atom are shifted down. The $LiA^{II}B^{III}$ and $A^{III}B^{II}Li$ structures come to be ground and obviously favorable isomers. The alternative $LiA^{III}B^{II}$ and $A^{II}B^{III}Li$ structures lie higher about 20–40 kcal/mol and are scarcely stable or not stable to a conversion into the ground isomer. The barriers between them do not exceed a few kcal/mol ($LiSiO^+$, $LiAlO$, $BSLi$) or disappear at all. In a latter case, the linear excited structures ($LiSiN$, $CSLi^+$, $CPLi$), appear to correspond to the top of the barrier, the $LiSiO^+$, $LiAlO$ and $BSLi$ only exist at low temperature conditions (in inert matrixes).

The migrational non-rigidity, typical for complexes with the light $A^{II}B^{II}$ ligands, becomes uncharacteristical for the $LiA^{II}B^{III}$ and $A^{III}B^{II}Li$ compounds. All the $A^{II}B^{III}$ and $A^{III}B^{II}$ ligands are coordinated to the cation by their light A^{II} or B^{II} atoms while other orientations are much less favorable in energy and scarcely (or not at all) stable kinetically.

It is not difficult to interpret these tendencies within the electrostatic model. When replacing one of the light A^{II} or B^{II} by its heavier A^{III} or B^{III} analog its electronegativity decreases, more for B and less for A. When passing from $A^{II}B^{II}$ to $A^{II}B^{III}$ or $A^{III}B^{II}$ the electron density is shifted from the heavier atom to the lighter one. As a result, A^{III} has a larger positive effective charge as compared to A^{II}, while B^{III} has a lesser negative charge as compared to B^{II}. This can be a major reason for essential destabilization of the $LiA^{III}B^{II}$, and of the $A^{II}B^{III}Li$ structure as compared to $A^{II}B^{II}Li$.

One can assume within the electrostatic model these tendencies to increase when replacing one of the ligand's atom along the subgroup to the fourth and lower periods of the Periodic Table. Potential energy surfaces of salts and ion-molecular complexes with the "ligands" BSe^-—BTe^-, CSe—CTe, NAs—NSb, CAs^-—CSb^-, GaO^-—InO^-, GeO—SnO—PbO, GeN^-—SnN^-—PbN^-, etc. have probably the single minimum corresponding to the linear coordination of the Li^+ cation to the light ligand's atom while the barrier (the alternative linear structure) increases when increasing the nuclear charge of the heavier atom along the subgroup. The bonding isomerism is not typical for the ligand of this type.

When the simultaneous replacing of both light A^{II} and B^{II} atoms by their heavier A^{III} and B^{III} analogs, the deformations of PES, which had opposite direction in the $LiA^{II}B^{III}$ and $A^{III}B^{II}Li$ systems, compensate each other because of decreases of electronegativity difference between the A^{III} and B^{III} atoms as compared with A^{II} and B^{II}. With this reason, the PES of the $LiA^{III}B^{III}$ systems are much closer to the PES of the similar light $LiA^{II}B^{II}$ species than to $LiA^{II}B^{III}$ or $A^{III}B^{II}Li$. Naturally, the compensation is not complete, the $LiA^{III}B^{III}$ isomer is destabilized more than $A^{III}B^{III}Li$, and the Li^+ is coordinated preferably to the electronegative B^{III} end. Difference in energies of the ground $A^{III}B^{III}Li$ and the excited $LiA^{III}B^{III}$ structures is decreased about twice when compared to the $A^{III}B^{II}Li$ systems, and does not exceed 15—20 kcal/mol. The barriers, separating the excited $LiA^{III}B^{III}$ isomer from the ground $A^{III}B^{III}Li$, are about twice higher, than for $A^{III}B^{II}Li$ (LiSiP is an exception), therefore the isomerism has to be more characteristical for $A^{III}B^{III}Li$ as compared to $A^{II}B^{III}Li$. At the same time the barriers do not exceed 5—10 kcal/mol, and the excited isomers $LiA^{III}B^{III}$ will probably exist at moderate and low (not too high) temperature conditions only.

PES of $SiSLi^+$ and $AlSLi$ are similar with PES of $COLi^=$ and $BOLi$ qualitatively. The quantitative difference is that the excited isomers for the sulfur derivatives lie higher essentially and are separated by the higher barrier. Moreover, the ground $AlSLi$ isomer is bent in contrast to the linear $BOLi$, although the inversion barrier of $AlSLi$ through the linear structure is small (1.5—2 kcal/mol) and the $\varphi(AlSLi)$ angle is an inversion non-rigid coordinate. The difference between $SiPLi$ and $CNLi$ is that the first molecule has the cyclic or triangle structure, similar to $NaCN$ and KCN [31—33] with Li^+ located about above the P atom. The inversion barrier here is small as well (~ 4 kcal/mol), and the $SiPLi$ isomer appears to be deformational and non-rigid, however slightly more rigid than $AlSLi$. PES of the $AlClLi^+$ and $AlFLi^+$ are similar; the quantitative difference is that the isomerization energy of $AlClLi^+$ is about 10 kcal/mol less, while the barrier is about 6 kcal/mol higher.

PES of LiN_2^+ and LiP_2^+ are most similar. The last ion is non-rigid to the migration of Li^+ around the homonuclear P_2 ligand. Its linear ground structure of the δ-complex is only 1.5 kcal/mol more profitable, than the cyclic π-complex structure corresponding to the top of barrier both for LiP_2^+ and for LiN_2^+. The barrier of LiP_2^+ is about twice as less as the LiN_2^+.

The barrier for LiSiP is small (~ 2 kcal/mol on the SCF level) and increases of the LiAlS, $LiSiS^+$ (~ 7 kcal/mol) and $LiAlCl^+$ (~ 17 kcal/mol). The bonding isomerism has to be more typical and ore clearly pronounced for the $A^{III}B^{III}Li$ systems than for $A^{III}B^{II}Li$ and $LiA^{II}B^{III}$. Both isomers can exist for AlSLi and $AlClLi^+$ and be detected in inert matrix isolation, LiSiP is scarcely stable to the conversion into SiPLi; for $LiSiS^+$ the barrier top is close to the dissociation $SiS + Li^+$ limit. Except for the migrational non-rigid LiP_2^+, the $A^{III}B^{III}Li$ systems remain "globally" rigid to the migration of the Li^+ around the space of the ligand $A^{III}B^{III}$ but "locally" non-rigid to the large amplitude vibrations in the vicinity of both isomers.

Returning back to the question about the transformation of the potential energy surfaces when replacing the A^{III} and B^{III} atoms by their heavier analogs from the forth, fifth and lower periods, within the electrostatic model one can wait for the following. When both ligand's atoms are from the same period, a difference in their electronegativity will decrease and the A—B bond becomes more and more covalent in a row $A^{III}B^{III}$—$A^{IV}B^{IV}$—$A^{VI}B^{VI}$. The bent or cyclic[a] ABLi isomers with the cation coordinated to the electronegative B end, remain to be ground but both isomers approach each other on the energy scale. A problem of barriers is more complicated, additional calculations are necessary. The $LiAs_2s$ +, $LiSb_2^+$ and $LiBi_2^+$ with homonuclear heavy ligands will probably be nonrigid to the quasi-free migration of the cation and the AA ligand relative to each other.

If we were to fix one atom from the third period, then the isomerization energy has to increase in rows of the SiS—GeS—SnS—PbS type with the electronegative B^{III} atom fixed, and to decrease in rows of the SiS—SiSe—SiTe type with the more electropositive A^{III} atom fixed.

Let us underline that the "heavier" diatomic AB^- anions (and probably the isoelectric neutral AB molecules), not containing elements of the second period, prefer to be coordinated as ligands of the π-type, with cation located above the A—B bond or the electronegative atom B (P, S, Al, Se, etc.), in contrast with carbonyl, nitrogenyl, cyanide and similar complexes are coordinated as δ-ligands, with linear M^{1+}—Ba fragments.

---

[a]Certainly, this classification is conventional: we will entitle "cyclic" and "bent" the structures with cation located above the A—B bond (all angles are less than 90°) and in the vicinity of one of the atoms (the valent angle with vertex at this atom is obtuse), correspondingly.

3.2 Comparison of PES of Rearrangements MAB → ABM (M = Li and Cu) and HAB → ABH

So far, we discussed the salt molecules and ion-molecular complexes with the Li^+ cation. There is a question however, how will the potential energy surfaces change when replacing Li^+ with monocharged cation of other metals. In {19,90–95] the PES'es of the CuAB → ABCu and LiAB → ABLi rearrangements were compared with the same AB ligands containing 10 valent electrons. One can see from Table 2 that related molecules and ion-molecular complexes of the monovalent copper and lithium are similar in relation not only to a shape and main features of PES, but of tendencies of the E_{isom} energies and h_{isom} barriers as well. As Li^+, Cu^+ is coordinated to the light $A^{II}B^{II}$ by both their ends forming the linear isomers, $CuA^{II}B^{II}$ and $A^{II}B^{II}Cu$. The first isomer is usually ground one and stabilized more (especially for $CuBO)$, that its lithium prototype $LiA^{II}B^{II}$, mainly because of the stronger $Cu—A^{II}$ bond as compared with $Li—A^{II}$. For the same reason, the isomerization energies for CuCP and $CuCS^+$ are higher than for LiCP and $LiCS^+$, while for AlOCu, on the contrary, lesser than for AlOLi. The ions LiP_2^+ and CuP_2^+ appear to be the migration ally non-rigid differing from LiN_2^+ and CuN_2^+ having essentially higher barriers. Keep in mind these results, one can assume the tendencies, described in detail in Section 3.1 for the Li^+ compounds, remain to be true, at least in a qualitative form, for similar Cu^+ compounds as well. The most serious difference is that the barriers, corresponding to the

bent or cyclic structures usually are

about twice as high, when compared to ,

and therefore compounds of Cu^+ have to be more rigid in migration of the cation around a ligand, than those of Li^+. Similar pictures can be proposed for compounds of Ag^+ and Au^+.

Non-empirical calculations of PES of isomerization for compounds of other transition metals are scarce. In [96] relative stability of isomers CuOLi → LiCuO and ScOLi → LiScO was discussed and found the MOM' structure with central oxygen atom and terminal both M and M' metal atoms to be much more profitable when compared to the alternative MM'O or M'MO containing direct metal—metal bonds. It was concluded [96], that the same MYM' structure to be ground one for many combinations of transition metals M and M' in oxides MOM', sulphides MSM' and another chalkogenides MYM'.

In [97–98] isomerization for $CuCO^+$ was studied, in [99] for $ScCO^+y$ and $CrCO^+$ ions, while in [100–110] only one of the alternative isomers was investigated.

It is interesting to make a systematical comparison of the PES of the LiAB → ABLi and HAB → ABH rearrangements of Li^+ and H^+ proton migration around the same AB ligands.

Isomerization of the 1,2-hydrogen shift in triatomic systems was studied theoretically in many papers (see for example, reviews [111,112] and references herein). In [112] a collection of data for HAB systems calculated within the MP3/DZHD + P approximation (similar to the approximation used in calculations of lithium compounds LiAB, see Section 3.1) is given. Corresponding profiles of potential energy curves along the minimal energy pathway of the HAB → ABH rearrangements with 10 valent electrons are plotted on Figure 3.

It can be seen from Figure 3, that replacing of cation Li^+ by proton H^+ in many cases does not change qualitatively a shape of PES, but stretches it along the energy coordinate resulting in an increase of isomerization energies E_{isom} and barriers h_{isom} by 2 to 3 times. The most pronounced similarity takes a place for ligands CP^-, SiN^-, SiP^-, AlS^-, N_2, NP and SiS. For others, one can mark some differences. For example, the most profitable isomer for the lithium cyanide is the isocyanide CNLi, while for the hydrogen analog—the cyanide one HCN (substitution Li^+ by H^+ reverses relative position of the isomers on the energy scale and sign of E_{isom} from -4 to +18 kcal/mol). The isomers LiBO and BOLi are quasi-degenerated in energy, while HBO is more profitable than HBO by ~ 40 kcal/mol. At the same time, both AlOLi and AlOH lie about 55 kcal/mol below as compared to LiAlO and HAlO, correspondingly. A very flat PES is typical for all isomers BOX, AlOX, COX^+ and $SiOX^+$ within the interval of the valent angle (with vertex at 0 atom) ~ 120 to 180°.

The excited BSH isomer has a bent structure. PES of BSLi has a shallow minima corresponding to the cyclic structure with Li^+ above the middle of the B—S bond and with very low inversion barrier. For HCS^+ picture is similar, while PES of the lithium thiocarbonyl has single $LiCS^+$ minimum, and $CSLi^+$ is a top of barrier. In LiP_2^+ the Li^+ cation is coordinated to P_2 molecules by one P atom forming a linear structure $C_{\infty v}$, while in HP_2^+ proton prefers to be coordinated to a middle of P—P bond forming a cyclic structure C_{2v}.

Isomers ABH with proton H^+ bonded with the electronegative B^{III} atom from the third period are usually of bent structure and much more rigid to deformation of $\varphi_{(HBA)}$: the inversion barriers via the linear structure for ABH are about 10 to 30 kcal/mol, while for ABLi they do not exceed a few kcal/mol or vanish at all. On the contrary, all isomers HAB and LiAB, whatever ground or excited ones, where Li^+ or proton is bonded with the electropositive A atom (no matter, from the second, third or lower period) are linear as a rule.

Comparing these results of Section 3.1, one can see the tendencies of transformation of PES, when substituting the light A^{II} and B^{II} atoms by their heavier A^{III} and B^{III} analogs, to be similar for LiAB and HAB with $N_{val} = 10$. When one of the ligand's atoms is from the second period and another from the third or heavier one, bond of proton or Li^+ with the light atom is much more preferable. The alternative structure with the bond or coordination to the heavy atom lies much higher in energy and corresponds either to the excited bent isomer or to a top of a barrier.

When both atoms are from the third or heavier period ($A^{III}B^{III}$, $A^{IV}B^{IV}$, etc.), then, because of partial compensation of two opposite tendencies (see Section 3.1), linear HAB (or LiAB) and bent ABH (ABLi) isomers comes to be more close in energy (for H^+ stronger, than for Li^+). As related to the isomerization energy, the $HA^{III}B^{III}$, $HA^{IV}B^{IV}$, etc. systems approach to their light $HA^{II}B^{II}$ analogs. Main difference is that for the light molecules the HAB structure is by ~ 15 to 35 kcal/mol more preferable than ABH, while for the heavier species the ABH structure with the more electronegative atom B protonated, is always more profitable by 5 to 20 kcal/mol. Discussing in the same way, as for LiAB compounds (Section 3.1), one can predict some qualitative tendencies for E_{isom} (HAB).

If one of the atoms A^{II} or B^{II} is from the second period and fixed, and another one is from the third and lower period, proton in the ground isomer will be bonded with the light atom, and energy E_{isom} and barrier h_{isom} (calculated off from the ground isomer) will increase when the nuclear charge of the heavier atom increases along its subgroup. Structures $HAsN^+, HSbN^+$, $HBiN^+$ (similar to HPN^+) and $HAsC, HSbC, HBiC$ (similar to HPC) correspond to the top of the barrier, their PES'es have single minimum with the H—N or H—C bonds, the isomerism is not characteristic here. Isomerization energy of ABH molecules and ions where N is a period's number and both heavy atoms are from the same period, has to decrease when increasing of N in parallel with the decreasing of the electronegativity difference between A^N and B^B. Bent $A^N B^N H$ isomer will be more favorable in energy with the $\varphi(ABH)$ angle decreasing when N increases. As for the barrier h_{isom}, it decreases relatively slowly in series $HSiO^+ - HSiS^+$, $HAlO - HAlS$, $HSiN - HSiP$, etc., and both isomers will exist probably for the $HA^N B^N$ systems with heavy A^N and B^N atoms.

Ion HN_2^+ has a ground linear structure $C_{\infty v}$ with a cyclic C_{2v} corresponding to the top of the barrier, while the cyclic structure of HP_2^+ with proton coordinated to the middle of the P—P bond is slightly more stable as compared to the linear $C_{\infty v}$ (top of barrier). The cyclic structure, probably, will be stabilized in rows of homonuclear ions of $HP_2^+ - HAs_2^+ - HSb_2^+ - HBi_2^+$ type.

As a whole, many qualitative tendencies of E_{isom} appear to be similar both for electropositive Li^+ and Cu^+ cation and for proton H^+, in spite of a much more covalent nature of the H—A and H—B bonds as compared to the Cu—A, Cu—B and the more with Li—A, Li—B.

The tendencies of $E_{isom}(HAB)$ for 1,2-hydrogen shift isomerization were analyzed in [25,111,113–117] within a valent bond model based on a covalent structure. Its logic can be followed with use of BOH → HBO and AlOH → HAlO rearrangements as examples. In the first (hydroxide) isomers, the monovalent electropositive A (boron or aluminum) atom is in a terminal position and bonded with oxygen by a single bond, while the latter isomers have trivalent B or Al in a central position and double B=O or Al=O bond. When migration of H from O to B(Al) the $\delta(O—H)$ bond is broken, $\delta(A—H)$ and $\pi(A—O)$ bonds are formed, and the atom B(Al) is promoted from the monovalent state into the trivalent one. Length R(AB) and overlapping population Q(AB) of double A=O bond are ~ 0.05 to 0.10 Å shorter and, correspondingly, 30 to 50% larger than those of single A—O bond. The bond O—H is usually stronger than A—H, and the valent state promotion energies of B or Al atoms are essential, ~ 60 to 100 kcal/mol and more. These energy expenditures have to be compensated by formation of $\pi(A—O)$ bond. Relative stability of the A—O—H and H—A=O isomers depends on which of these two factors is dominating for given electropositive A atom. Very strong $\pi(A—O)$ results into the ground H—A=O isomer with multiple A=O bonds. If $\pi(A—O)$ is weak, the hydroxide A—O—H isomer becomes to be close in energy to H—A=O or even more profitable than H—A=O.

It is well known that $2p\pi—2p\pi$-bonds between light A^{II} and B^{II} atoms from the second period are very strong and comparable (or even stronger) with corresponding $2p\delta—2p\delta$-bonds. Therefore, the most stable structures of compounds of the light elements are those where the most number of the strong π-bonds and highest valencies of A atoms are realized. The isomers of type H—B=O, H—C≡N, H—C=O$^+$, H—N=O$^+$ appear to be not only ground, but much more preferable in energy than the alternative isomers with a less number of π-bonds. Opposite, energies of $3p\pi—2p\pi$, $3p\pi—3p\pi$, $4p\pi—2p\pi$, etc. for the A^{III}, A^{IV} and heavier atoms from the third and lower periods decrease sharply as compared not only with $2p\pi—2p\pi$ prototypes, but with energies of the corresponding δ-bonds.

Therefore, when changing the A atom alone the subgroup down the Periodic Table, the isomers with multiple bonds will be usually destabilized much stronger than isomers with ordinary δ-bonds. As a result, the alternative isomers sharply approach each other on the energy scale (as compared to the relative compounds of the light elements, including an inversion of their relative stability and changing of the ground isomer. The most preferable become the structures with more numbers of stronger δ-bonds and with less numbers of weak π-bonds: Al—OH, Si—OH$^+$, Si—OH, etc. with the valent non-saturated heavy A^{III}, A^{IV} and A^V atoms having a lone electron pair. In cases of this kind, a periodicity rule is violated between second and third periods but preserved below from the third to IVth-VIth periods.

Surely, the model [113–116] is of qualitative character, but it describes satisfactory relative position of isomers for several hundred of covalent molecules from the simplest HAB to the more complicated Si_6H_6 or P_2O_x with x = 2.5 (see [25,117] in more detail). As mentioned above, the model [113–116] operates with covalent structures of the HAB and ABH isomers.

For high-polar salt molecules and ion-molecular complexes of metal cations, similar tendencies of E_{isom} (LiAB) can be described no less satisfactory within an electrostatic model, operating with ionic structures Li^+,AB^- and AB^-,Li^+. Tendencies of other molecular characteristics for proton and metal cations are similar as well: when migration of M^+ from electropositive A atom to electronegative B, R(AB) distance is always elongated, R(AB) force constant and overlapping population Q(AB) are decreased, when passing from HAB to ABH. The difference is the quantitative character. In covalent HAB molecules and ions $\pi(A{-}B)$ bond is broken, and these changes are large, while in CuAB and LiAB salts the changes are ~ 1.5 to 2 times less, and it is more sufficient here to talk about not breaking the A—B bond but rather polarizing the AB ligand by cation. The polarization can be essential or minor, depending on the polarizing ability of the cation and the polarizability of the ligand. For Cu^+ it is larger than for Li^+, and decreases to Na^+, K^+ and Rb^+. Its manifestation is weaker for the most rigid and strong CO, N_2 or CN^- ligands but increases sharply to CS, CP^-, SiO, BO^- or BS^-, and reaches maxima for BF, BCl, AlF and AlCl. For example, when going from $LiAlCl^+$ to $AlClLi^+$, the Al—Cl bond becomes weak and transforms from the two-center to the three-center bridged $Al{-}Cl{-}Li^+$; the R(Al—Cl) distance is elongated by ~ 0.3 Å. For the proton, the elongation is even larger, because the Al—Cl bond in the $H{-}Al{-}Cl^+$ structure transforms into ion-dipole interaction in the $Al^+...Cl{-}H$ structure.

In real compounds under study, the bonds of cation or protons with ligands AB are neither pure electrostatic nor pure covalent, both components are mixed in various proportions. Keep in mind that the electrostatic model and the covalent model of $\pi(A{-}B)$ bonds [113–116] give a similar picture for both extreme cases, one can hope the tendencies above to remain true for other M^+ cation with electronegativity intermediate between alkali metals and protons. When increasing the electronegativity of M, PES will be "stretched out" along the energy coordinate, approaching to PES of the 1,2-hydrogen shift; if electronegativity of M decreases, PES will be flattened, approaching to that of the LiAB salts.[a] The same conclusion can be obtained within the covalent model [113–116] taking into account the dependence of the isomerization energy E_{isom} (XAB) on nature of migrating atom or group X, with the use of an approximate additive schema and bond energies E(A—X) and E(B—X): the first bond (with the electropositive A atom) is usually to

---

[a]It is true not only for monatomic metal cations M^+, but for more complicated MgH^+, AlH_2^+, SiH_3^+, GeH_3^+ and similar polyatomic ions, migrating around AB ligand as a single structural group.

strengthen and the second one is to weaken with the increase of the electronegativity of X.

Discussing similarity of PES for LiAB and HAB, we minded the most general tendencies and essential energetic effects. Due to the approximate character of calculations, it is risky sometimes to analyze "fine" effects of one or two kcal/mol, although even so small differences can be interesting for non-rigid species with a very flat PES, when localizing the global minimum and corresponding ground structure. For example, PES of the cyanide BeCN, MgCN radicals and BeCN$^+$ ion is similar to that of LiCN, but a barrier (bent structure II, see Figure 1) separating the ground isocyanide V and excited cyanide I isomers, decreases in a row BeCN (12 kcal/mol)—BeCN$^+$ (8 kcal/mol)—MgCN (2 kcal/mol).[b] PES of more heavier CaCN and BaCN analogs has a single isocyanide minima V with the structure I being at the top of the barrier about 7 to 8 kcal/mol [120]. In these cases even relatively small perturbations (replacing of the light Li and Be cations by the heavier alkali and alkali-earth metals) can be enough for the deformations of PES and equilibrium (non-rigid) valent angles.

3.3 Role of the Charge of Cation and its Chemical Environment

Let us consider, how will a shape change PES and what are the tendencies of E_{isom} and h_{isom}, if to change 1) a charge q of M^{q+} cation and its position in the Periodic Table, 2) a chemical environment of the cation. The most detailed calculations of PES were performed for cyanide [91,120], carbonyl [91,120–125] and nitrogenyl [122,124,125] complexes. Their results are given in Tables 3 and 4. Alternative structures corresponding to special points of PES are drawn on Figure 4.

It can be seen from Tables 3 and 4 that the shape of PES of $M^+ \cdot CO \rightarrow M^{q+} \cdot OC$ and $XM \cdot CO \rightarrow XM \cdot OC$ rearrangements is preserved qualitatively for all ion-molecular complexes $M^{q+} \cdot CO$ with mono-, di- and tricharged cations and for neutral $XLi \cdot CO$ and $XCu \cdot CO$ systems, with X "acido"-substitute, varying from the least electronegative F to the least electropositive Li. Differences are of quantitative character and follow the tendencies discussed below.

Energy D(M—CO) of the metal—carbonyl bond increases sharply with the increase of the cation charge q (in kcal/mol: 10 for HLiCO, 17 for Li$^+ \cdot$ CO, 100 for Be$^{2+} \cdot$ CO and 144 for Al$^{3+} \cdot$ CO), but decreases quickly with changes of the cation along the subgroup down the Periodic Table (49 for Mg$^{2+} \cdot$ CO and 23 for Ca$^{2+} \cdot$ CO). Tendencies of barrier h_{isom} are similar but pronounced more smoothly (5 for HLi $\cdot$ CO, 10 for Li$^+ \cdot$ CO, 32 for Be$^{2+} \cdot$

---

[b]On the SCF level; an electron correlation makes them probably slightly lower.

CO and 45 for $Al^{3+} \cdot CO$). As for relative energies of isomers, their qualitative trends remain the same, but in much more narrow limits as compared to h_{isom}. Energy E_{isom} is maximal for polycharged cations $Be^{2+} \cdot CO$ and $Mg^{2+} \cdot CO$ (10 to 12 kcal/mol) while for other ions mentioned in Table 3, including $Ca^{2+} \cdot CO$, it does not exceed a few kcal/mol. Its values are even less and their dependence on the X substitute is weaker in the neutral XLi $\cdot$ CO systems: when replacing H by F E_{isom} increases about 1 kcal/mol only. One can suppose energetic closeness of ground MCO and excited MOC isomers within a few kcal/mol to be typical for many carbonyl complexes with cations of electropositive metals, especially of heavy ones from the fourth and lower periods. Inductive effect, related with the changing of electronegativity of X, influences on E_{isom} rather weakly. When CO approaching Li in LiH_2^- anion by both C and O ends, its PES is flat with shallow minima ~ 1 to 2 kcal/mol, therefore negative charged complexes of the $X_2Li^- \cdot CO$ type have to be scarcely stable to the breaking off of the CO group.

Potential barriers are large for polycharged light cations and make it impossible for isomerization under intramolecular mechanism, although for heavy metals the decrease is strong. For monocharged cations (alkali and alkali earth metals, Al^+ and its analogs) and for neutral XLi $\cdot$ CO or XCu $\cdot$ CO species h_{isom} does not exceed 5 to 10 kcal/mol, but lies close to a dissociation limit. The overcoming of the barrier here can be accompanied by dissociative processes. In weakly bonded negative ions like $X_2Li^- \cdot CO$ the dissociation limit is very low, and the isomerization, if possible, will go under the dissociative/associative mechanism. For Li^+CO h_{isom} is about 10 kcal/mol, for XLi $\cdot$ CO it decreases with the increase of the electronegativity of X (7 kcal/mol for FLi $\cdot$ CO, 5 for HLi $\cdot$ CO and 3 for LiLi $\cdot$ CO). The most non-rigid are carbonyls of heavy elements Rb^+, Sr^{2+}, Ga^+, etc. For $Cu^+ \cdot CO$ the barrier is higher than for Li^+CO. Carbonyls of transition metals will be more rigid to the isomerization. Similar picture is received for cyanide and nitrogenyl complexes (see Table 3).

Until now, we have discussed the simplest fragments and complexes with one CO group. However, we question how relative energies of MCO and MOC orientations will change in more complicated carbonyls, containing several CO groups, and what is their mutual influence. In [122] the SCF and MP4/6-31G calculations of $Li^+(CO)_m$ ions with m varying from 1 to 4, were carried out. It was demonstrated that isomerization energy per one CO group is even less sensitive to the number of carbonyl groups, that to electronegativity, in above mentioned XLiCO and XCuCO neutral complexes. Barrier h_{isom} depends on m relatively weakly as well, the strongest dependence is manifestated for the energy $D(Li^+—CO)$. The intramolecular mechanism of Lis + $\cdot$ CO → $Li^+ \cdot$ CO isomerization in $Li^+(CO)_m$ ions can be realized only for small m (if possible at all). At large m passing over the barrier can be less advantageous than the dissociation of the cation—carbonyl bond. When increasing the positive charge q both intramolecular and dissociation mechanisms are hindered because of the sharp increase of $E(M^{q+}-CO)$ and h_{isom}.

A problem is if the results above obtained for simplest $M^{q+} \cdot CO$ or XM $\cdot$ CO fragments can be transferrable into "real" carbonyl complexes containing several CO (or other neutral donor) groups and acido—ligands X. One can expect the best transferability in iso-charged series with various number of ligands but the constant oxidation number of central M^{q+} cation, by analogy with the discussed $Li^+(CO)_m$ ions. In acido-complexes energy E_{isom} has to increase slowly and h_{isom} and $D(M^{q+}-CO)$ several times faster when increasing the electronegativity of X. When increasing the outer positive charge of the complex ion and of the oxidation number of a central cation, the tendencies will be pronounced more distinctly.

A reason for this dependence is mainly an electrostatic nature of $M^{q+}-CO$ bond in the carbonyl complexes of nontransition metal cations in accordance with an energy decomposition analysis by Morokuma [119]. The carbonyl group plays a role of δ-donor here and is weakly polarized in the coordination sphere. All δ- and π-bonds remain strongly localized within M—CO and M—X fragments, and their non-additive components of energy are not large. One can suppose a similar picture for nitrogenyls, cyanides and related complexes with strong and hard ligands.

A situation is more complicated for a less strong and softer ligands CS, CP⁻, NP etc. If for "rigid" carbonyls and cyanides E_{isom} usually lies within narrow limits of about 10 kcal/mol, for "less rigid" thiocarbonyl and similar ligands E_{isom} depends on a charge q of a central M^{q+} cations much more stronger. For example (See Table 3), when passing from Li^+ to Be^{2+} a preference of the $M^{q+} \cdot CS$ structure against $M^{q+} \cdot SC$ increases from ~ 30 to ~ 90 kcal/mol, and that of the $M^{q+}MP$ structure against $M^{q+} \cdot PN$—from ~ 50 to ~ 140 kcal/mol, i.e., about three times. the linear $M^{q+} \cdot SC$ and $M^{q+} \cdot$ PN are saddle points of PES and not stable to conversion into $M^{q+} \cdot CS$ and $M^q \cdot NP$ minima. Coordination by S and P atoms, linear or bent, is not profitable. For fragments with CP⁻, BO⁻, AlO⁻ and similar "ligands" the same or even stronger dependence on q is typical. When going from Be^{2+} to Mg^{2+}, Sr^{2+} and heavier alkali-earth cations, the dependence becomes much smoother in parallel with the dropping polarizing ability of the M^{q+} cation.

As it was mentioned above, coordinations of light ligands CO, N_2, CN⁻ and BO⁻ to M^{q+} cations by both their A^{II} and B^{II} ends are close in energy. If one atom is from the second period and another one—from the third or lower period, coordination by the lighter atom is much more preferable and in many cases that is a single one. If both ligand's atoms correspond to the third or heavier period, coordination by both ends becomes possible. Keep in mind, the results of this section, one can suppose the picture does not change qualitatively when passing from isolated MAM fragments to complexes (or to another active centers on a surface) with a central M^{q+} cation surrounded by other ligands in the coordination sphere or by neighbor atoms. If the oxidation number of M does not change (other ligands are neutral), a ligand's mutual influence will not be strong enough to transform a qualitative shape of PES corresponding to the isomerization of the M^q—AB fragment. PES will remain to be relatively flat for CO, CN⁻ and N_2, ligands CS, CP⁻ and CP⁻ as a rule will be coordinated by their light atom, etc.

These resurls are related mainly to complexes of nontransition metals. From Table 4B a similar picture for Cu^+AB and $Zn^{2+}AB$ can be seen. It will probably be preserved for other cations with the closed d^{10} (and may be, half-closed d^5) shell, but can be more complicated for open d^n shells where n ≠ 5 and 10 because of dative interactions.

In Table 5, optimized geometry and energy characteristics are given for special points of PES of $(H_2O \cdot Li^+) \cdot AB \rightarrow (H_2O \cdot Li^+) \cdot BA$ rearrangements with Li^+ solvated by one water molecule. It can be seen that H_2O molecule weakly perturbs a LiAB fragment: R(LiA) and R(LiB) distances are elongated by a several hundredth of Å, both in linear and bent structures (the last one corresponds to a top of barrier, its angle φ(LiOA) decreases by a few degrees), R(LiO) varies in narrow limits 1.80–1.85 Å, redistribution of electron density (atomic effective charges and overlap populations) between cation and ligands AB is small—a few hundredths of an electron. Special points of the PES approach each other on the energy scale by 1–4 kcal/mol, and PES'es come to be more flat as compared to the isolated LiAB fragments. A weakening of the Li^+—AB bonds is more essential, their energies decrease by 2–5 kcal/mol for ion-molecular complexes with neutral ligands (CO,NP) and by 10–15 kcal/mol for salts with BO⁻, CN⁻, CN⁻, SiN⁻ and SiP⁻ anions. When increasing the number n of water molecules inside the coordination sphere, these effects are accumulative $LiA^{II}B^{II}$ and $LiA^{III}B^{III}$ fragments become more non-rigid in migration, especially with anions, and the Li^+—AB bond is weakening the most. When saturation of coordination sphere by H_2O molecules, the cation—ligand distance becomes the non-rigid coordinate and elongates by tenths of angström or more without a tangible change of energy. Actually, the ligand is pushed out from the first coordinate sphere to the second one (in case of the AB⁻ anion a "contact pair" divided by a hydrate shell is formed) where rotations or large amplitude movements of the cation and anion relative to each other are essentially less hindered, and for the light $A^{II}B^{II}$ ligands they appear to be quasi-free. At the same time, the ligands like CP⁻, CS and NP remain coordinated by the light atom, while orientation by a heavy end is scarcely stable (it is possible, in principle, in special situations with favorable steric factors, but that topic is beyond our consideration). At large, R migration can be free.

3.4 Fragments LiAB of Diatomic Ligands with 12 Valent Electrons

In a literature there are non-empirical calculations of PES of isomerization for molecules LiNO [131,133], LiPO [133,134], LiNS and LiPS [133] in their lowest singlet and triplet electronic states. Results of calculations performed in [133] are given in Table 6, profiles of their PES[a] are plotted on Figure 5.

Contrary to previous systems with $N_{val} = 10$ in a singlet state all the species with 12 valent electrons under consideration have a cyclic ground structure III with the bridged (not terminal) cation M, located above the A—B bond. Other structures, including I, II and V, are less profitable in energy with the last one (the linear V, where Li^+ is coordinated to the S atom corresponding to a top of a barrier. The difference in the energy of I, III and V is large enough to regard molecules like LiNO as locally-nonrigid to movements of cations along the A—B bond within large limits of angle φ in the vicinity of III. Migration of the cation around all of the space of the anion in not probable up to high temperatures.

One can see from Table 6 that qualitative tendencies of relative energies of I, III and V structures, as well as angle φ in the equilibrium ground structure III, are adequately described within the simple electrostatic model. For example, when replacing N by P and going from LiNO to LiPO, the negative charge on O increases and on P decreases resulting in the stabilization of V and the destabilization of I as compared to III. In the cyclic structure the cation is shifted to atom O area with the increase of $\varphi(NOLi)$. On the contrary, if we should replace O by S and pass from LiNO to LiNS, the negative charge increases on N and decreases on the chalkogen atom, resulting in the decrease of φ (because of the shift of the cation to the N atom area), stabilization of I and the destabilization of V comparatively with III. When both light atoms are replaced (from LiNO to LiPS) these two opposite tendencies are compensated in a certain degree, and differences in relative energies and φ for LiNO and LiPS appear to be least. Using the same model one can think the φ angle to decrease while the I and III structures approach each other up to the inversion of their relative position on the energy scale, and V is destabilized in series of the LiNS—LiNSe—LiNTe type. In rows of the LiPO—LiAsO—LiSbO—LiBiO type, φ has to increase, the III and V structures approach each other (again, up to their inversion) and I is destabilized.

---

[a]Anions AB^- with $N_{val} = 12$ have close-lying triplet and singlet states. Calculations of their salts LiAB and data of Table 6 and Figure 5 and be slightly more approximate as compared to the systems having 10 or 14 valent electrons with closed shells. On a section of PES between linear I and cyclic III structure within $\varphi \sim 30-60°$, two close-lying singlet terms of the electronic $(a')^2$ and $(a'')^2$ configurations can interact and be mixed producing shallow minima. The use of two-determinate reference wave functions, as it was done in [135] for LiSiH molecule, gives a more reliable result, but quantitative improvements are not too large (see [133] for more detail).

Within the SCF approximation, for all molecules with $N_{val} = 12$ their PES of the lowest triplet state (T) lies ~ 15–35 kcal/mol below as compared to the singlet (S). The S, T-splitting is large for the linear I and V structures and is minimal in the vicinity of III. The electron correlation energy will decrease this difference, and for the cyclic LiPO and LiPS triplet and singlet can be close in energy. For LiNO(T) PES is flat and typical for non-rigid systems. Its cyclic isomer is ground, the linear I and V isomers are only 3–4 kcal/mol above, and the barriers II and IV are low. In accordance with the electrostatic reason, LiPO(T) has a single minimum V, and I is the top of the high barrier. In opposite, LiNS(T) has a global minimum I with V indicating a top of the barrier, and a very flat section of PES with the slightly tangible local minimum in the vicinity of III. LiPS(T), similar to LiNO(T), has the single cyclic structure, but I and are tops of the barriers, contrary to LiNO(T).

3.5 Complexes LiAB with 14 Valent Electrons

Potential energy surfaces for the series of the LiOF, LiOCl, LiSF, LiSCl molecules and of the LiF_2^+, $LiFCl^+$ and $LiCl_2^+$ ions were calculated in [136] (LiF_2^+ and $LiFCl^+$ has been studied earlier in [137–139]). Their profiles are drawn on Figure 6, geometry and energy characteristics are collected in Table 7.

As one sees from the comparison of data of Tables 6 and 7, as well as Figure 5 and 6, molecules LiAB with 12 and 14 valent electrons are similar in the sense that they all have the cyclic ground structure III with Li^+ located above the A—B bond. A difference is that for $N_{val} = 14$ the cyclic structure corresponds to the single minimum of PES while both linear I and V structures to tops of the inversion barriers at the A and B atoms. These barriers are large enough (~ 20 kcal/mol and higher), and the migration of Li^+ around the whole space of the ligand becomes hindered or impossible. AT the same time, all the systems with $N_{val} = 14$ are "locally" non-rigid to large amplitude shifts of Li^+ along the A—B bond around the equilibrium III structure. LiOCl is the exception: its inversion barrier through the linear Li—O—Cl configuration is small, and therefore LiOCl can be non-rigid to the limited migration of Li^+ not only along the O—Cl bond but around the oxygen atom as well, avoiding of vicinity of the Cl atom because of a very high (~ 50 kcal/mol) inversion barrier via the linear O—Cl—Li structure (non-stable to dissociation O + ClLi).

As in the case of $N_{val} = 12$, the electrostatic model adequately describes the qualitative tendencies of the relative energies of I, III, V structures and of the angle φ for the equilibrium III configuration. From LiOF to LiOCl I is stabilized and V destabilized as compared with III, in the latter Li^+ is shifted closer to the O atom, and φ decreases in parallel with the increase of the negative effective charge of O and the decreasing of that on the halogen atom. Contrarily, from LiOF to LiSF I is destabilized and V stabilized relative to III with the shifting of Li^+ closer to the F atom in III and with the increase of φ When substituting both O by S and F by Cl, the opposite trends are compensated essentially and the differences in the relative energies and φ values between LiOF and LiSCl is much less.

For an electrostatic reason, I and III will approach each other in energy up to quasi-degeneration or inversion with a parallel V destabilization in the series LiOF—LiOCl—LiOBr—LiO. The hypobromides and hypoioids are supposed to be more non-rigid toward inversion. In the series like LiOF—LiSF—LiSeF—LiTeF I will be destabilized with V approaching to III in energy and φ increasing in III.

A trend to coordination by bond is still more pronounced for the isoelectronic ion-molecular complexes, having a very flat PES, especially the non-rigid LiF_2^+ and $LiCl_2^+$ ions with homonuclear ligands (see Figure 6). Within the wide interval of $\varphi = 60-120°$, all changes in energy are so small, that the calculated values of their equilibrium geometry parameters can be as strongly dependent on the computational approximation, as for the NaCN and KCN cyanides (see Section 2 and 3.1). In the SCF approximation with the $3\text{-}21G^*$ [137] and $6\text{-}31G^*$ [138] basis sets the bent structure of LiF_2^+ with $\varphi(LiFF) \sim 110°$ is slightly more preferable, while with the use of the $3\text{-}21G^*$ basis the isosceles triangle C_{2v} structure appears to lie slightly lower for LiF_2^+ and $LiCl_2^+$ [136]. For the final answer the using of more complete basis sets like DZHD + 2P or better is necessary. The $LiFCl^+$ ion has on its PES a single minimum (the quasilinear IV structure) with Li^+ coordinated to the electronegative F atom and a very low inversion barrier via the linear Li^+—F—Cl(V). Here Li^+ can migrate quasi-free along the F—Cl bond and around the F atom avoiding the area around Cl (the barrier about ~ 17 kcal/mol). In a similar way, the linear I structure of LiF_2^+ and $LiCl_2^+$ corresponds to a top of barrier about 6 and 10 kcal/mol, their global migrational non-rigidity would be possible at high temperature conditions only.

As for $N_{val} = 12$, PES'es of the LiAB $\rightarrow$ ABLi and HAB $\rightarrow$ ABH rearrangements with the same AB ligands having 14 valent electrons, differ drastically. In accordance with calculations and experimental data (see [111] and references herein), a global minimum of the HOF, HOCl and HSF molecules corresponds to the bent structure II, a local minimum—to the bent IV and a maximum—to the cyclic III. The inversion barriers (the linear I and V structures) are high.

It is difficult to rationalize a bent structure of species with 12 and 14 valent electrons within an idea of AO's hybridization, or Walsh diagrams or the Gillespy—Newholm model. An attempt was made in [137] to explain the peculiarity of LiF_2^+ structure by a donor-acceptor interaction of the electron pair on the antibonding $\pi^*(F_2)$—MO and the vacant 2p(Li)—AO.

One can think this interaction to increase $2p\pi(Li)$—AO population in the cyclic III structure, split the π^*-orbital energy into two components (a_1 in a plane of the cycle and b_2 perpendicular to the plane), to elongate the A—B bond and increase its overlap population Q(A—B). Calculations [136] demonstrate that all these effects really take place, but they seem to be relatively small[a] to believe that only (or mainly) they determine so drastic difference of PES'es for the systems with N_{val} = 10, on one hand, and with N_{val} = 12 and 14, on another. For example, the population of the $2p\pi(Li)$— AO (in a plane of the cycle) is about 0.1 e for LiOF, LiOCl and about twice more for LiSF, LiSCl. The splitting of a_1 and b_2 orbital energies never exceeded 0.4–0.5 eV. The R(AB) distance shortens (as compared with the isolated AB⁻ anion) by ~ 0.1 Å for LiCl, does not change for LiOF and LiSCl and lengthens by ~ 0.05 Å for LiSF. The overlap population Q(A—B) changes slightly, and it is difficult to follow its correlation with predictions of the model [137]. In the $LiF_2{}^+$, $LiFCl^+$ and $LiCl_2{}^+$ these effects are even less than in the salts.

3.6 Complexes M[ABC] of Triatomic Ligands ABC with 16 Valent Electrons

3.6.1. <u>Chalkogenides YAY′(Y,Y′ = 0 and S, A = B⁻, Al⁻, C and Si, etc.)</u>. Let us start from the light symmetric ligands AY_2 = $BO_2{}^-$ and CO_2 with N_{val} = 16. PES'es of their salt molecules and ions with the Li⁺ and Cu⁺ cations were calculated in [141–142]. In the isolated state these ligands, except the ground linear Y = A = Y structure, can have a local minimum of the excited cyclic structure   having the Y—Y bond and a lone electron pair on the A atom, but the latter lies high (by ~ 40 kcal/mol for the $AlS_2{}^-$ [141] and SiS_2 [143] disulphides; for oxides for Al and Si as well as for oxides and sulphides of B and C the cycle is supposed to be less stable). Therefore, here we limit ourselves by consideration of the I—VI structures, plotted on Figure 7 and corresponding to special points of PES along the MEP of the Li⁺ cation migration around the YAY′ ligand with the linear (or bent, but not cyclic) skeleton. The profiles of the PES'es are drawn on Figure 8, the optimized geometry and energy data are collected in Table 8.

One sees from Figure 8 and Table 8, that the contrary to the diatomics BO⁻ and CO, the isomerism is not typical for $LiBO_2$ and $LiCO_2{}^+$ with the triatomic ligands. Their PES has two equivalent minima I and V for the linear[a] structure Li—O—A—O with Li⁺ coordinated to one O atom. The

---

a

[a]The SCF and MP2 calculations with the 3-21G* and DZHD + P basis sets [141,144,145] testify the linear Li—O—B—O(I) structure, but results of the electron diffraction studied at high temperatures [68–70] and the IR spectra in inert matrix isolation [59–60] were interpreted in favor of the bent II structure with $\varphi(LiO_bB)$ ~ 100°. When sophistication of computation

cyclic C_{2v}(III) structure with bidentately coordinate Li^+ is a top of the migration barrier. The latter is about twice higher than in the triatomic LiBO and $LiCO^+$ species with a lone electron pair of B and C, having rather large affinity to Li^+ [139], and with the lower (about twice in comparison with $LiBO_2$ and $LiCO_2^+$) positive effective charge on the central A atom. The complexes of BO_2^- and CO_2 are essentially more rigid than these of BO^- and CO to the migration of cations and anions relative to each other. The ion-molecular complexes as usual are more rigid than the salt molecules. The migration barrier of $LiCO_2^+$ is close to its dissociation limit.

Similar to the $LiBO_2$ and $LiCO_2^+$, PES'es of $LiAlO_2$ and $LiSiO_2^+$ have a single minimum of the linear structure I. A difference is that, because of the sharp increase of the positive charge of the central atom, when replacing A^{II} by its analog A^{III}, a section of PES in the vicinity of the cyclic III is strongly destabilized and the barrier for the $LiAlO_2$ increases about twice against of $LiBO_2$ and $LiCO_2^+$. If to continue the tendency in the series A^{III}—A^{IV}—A^{V}—A^{VI}, all the oxides $LiAO_2$ with 16 valent electrons have the ground linear structure I and the barrier increases along the rows $LiSiO_2^+$—$LiGeO_2^+$—$LiPBO_2^+$, $LiAlO_2$—$LiGaO_2$—$LiInO_2$, etc.

When substituting both O atoms by S (A is fixed), the trends are reversed. From oxides $LiAO_2$ to sulphides $LiAS_2$ the positive effective charge of the A atom sharply decreases, and the PES in vicinity of III is strongly stabilized. For the $LiCS_2^+$ ion the cyclic and linear structures become close in energy, and for $LiBS_2$ they are inverted: III is a ground configuration and I is a top of the barrier. A global minimum of PES of $LiCS_2^+$ corresponds to the bent V with $\varphi(LiS_bC) \sim 125°$, while I and III are tops of the barriers. Extrapolating the trend, one can wait the cyclic III to be the ground or close to ground structure for selenides and tellurides $LiBSe_2$, $LiBTe_2$, $LiCSe_2^+$. In the series S—Se—Te II and III will be stabilized as compared to I. The isomerism is not typical for these compounds.

If to replace now both central atom A^{II} by A^{III} and O by S simultaneously, the opposite trends above are partially compensated (not completely: effect of the replacing of O by S is stronger) PES of $LiAlS_2$ and $LiSiS_2^+$ are

---

approximation with the use of more flexible DZHD + 2P (the double-zeta polarization functions [141]) and the MP3 and MP4 perturbation theory, the linear structure remains to be slightly more preferable, but the relative energy of the linear and bent structures decreases and in principle may be diminished in more accurate calculations. This PES is flat within the range of $\varphi(LiO_bB)$ $\sim 100–180°$, and on the semiquantitative level of the present consideration it is reasonable to restrict ourself by the conclusion on a quasi-linearity of LiBO and its deformational non-rigidity. For the final obtaining of the equilibrium $\varphi_e(LiO_bB)$ value, the more accurate approaches are necessary as well as a recalculation of the effective r_g, φ_g data measured experimentally to their equilibrium r_e, φ_e values (see [25,26,141]).

intermediate between those of the $LiBO_2$, $LiCO_2^+$ oxides and the $LiBS_2$, $LiCS_2^+$ sulphides of the light element, as was mentioned for the complexes with diatomic ligands (see Section 3.1). A difference is that the oxides have the linear structure I while for the sulphides, selenides, etc. the bent II is preferable with the inversion barrier I increasing from S to Se and Te. One sees from Figure 8 that $LiAlS_2$ has very flat PES and can be regarded as a migrationally non-rigid molecule.

At last, when going from oxides $LiAO_2$ to oxisulphides LiOAlS, the linear Li—O—A—S structure I with Li^+ coordinated to O appears to be obviously most stable, and the alternative O—A—S—Li in salts is scarcely stable to the conversion into I. The isomerism is not a characteristic for the salt molecules but can be probably observed for the ion-molecular complexes like $LiOCS^+$, where the barrier is not negligible.

It is seen from Table 8 that as in the complexes with diatomic ligands (Section 3.2), a shape of PES of the $MYAY'$ molecules and ions does not change qualitatively from Li^+, to Cu^+, although compounds of Cu^+ are more rigid than those of Li^+.

The "mixed" anions like OBS⁻ are coordinated to Li^+ and Cu^+ by their light O end, the orientation to S is not favorable. If replacing Li^+ and Cu^+ by proton, both bent HOAS and OASH isomers are stable and separated by the high barrier in the vicinity of the cyclic $\overset{O}{\underset{S}{\diagup}}A - H$ structure [147]. It is worth it to underline that the proton affinity values of the O and S end of OCS and OBS⁻ are close within a few kcal/mol. In the series OCS—OCSe—OCTe and OBS⁻—OBSe⁻—OBTe⁻ a preference of the protonation of the heavier chalkogen, against oxygen is increased in contrast with the related Li^+ and Cu^+ compounds where the ligands are always coordinated by O and the alternative structure is destabilized[a] from $OCSLi^+$ to $OCTeLi^+$. A reason is [117] that when protonating, the π-bond of the attached chalkogen with the central A atom is broken. As it was said in Section 3.2, energy of the $2p\pi_c$—$3p\pi_s$ bond is much less when compared to $2p\pi_c$—$2p\pi_O$ and decreases more to $2p\pi_c$—$4p\pi_{Se}$. Therefore, the protonation of the heavier chalkogen atom is more profitable than that of oxygen, and the proton affinity increased in the series CO_2—OCS—OCSe—OCTe in accordance with the CIPSI/DZHD + P calculations [148]. It is true as well for protonation of OBS^+ and related systems containing O and the central A^{II} atom from the second period. If to replace A^{II} by its heavy analog A^{III} (OAlS⁻, OSiS, etc.) the tendency is reversed. Difference in energies of the $3p\pi_A$—$2p\pi_O$ (Al—O or Si—O) and $3p\pi_A$—$3p\pi_S$ (Al—S or Si—S) bonds decreases sharply as compared to π-bonds C—O and Si—O, B—O and B—S. Therefore for the

---

[a]It is true for the diatomic CO and CS ligands as well. In CO the Li^+ affinity of the O atom is about 15 kcal/mol and in CS the Li^+ affinity of S does not exceed a few kcal/mol, while the proton affinity of O in CO is about 13 kcal/mol less than that of S in CS.

related mixed chalkogenides of Al, Si, P and their heavier analogs, the protonation (as well as coordination by cations like Li^+) of more electronegative oxygen is more advantageous, in accordance with calculations (see Table 8).

3.6.2. <u>Cyanates and thiocyanates</u>. In accordance with calculations [149–151], anions NCO^- and NCS^- have two linear isomers separated by high barriers (see Figure 9,10). In the ground cyanate and thiocyanate isomer I a chalkogen atom is bonded with C, the excited III (CNO^- and CNS^-) lie by ~ 65 and 40 kcal/mol higher. In [149] a shallow local minimum corresponding to the oxaziryl cycle II was found, but the latter, if it is able to exist, will be scarcely stable to the conversion into I because of the barrier between II and I is small (about 2 kcal/mol at the MP4/6-31+G* level). When scanning PES of the NCS^- → CNS^- rearrangement in the SCF/3-21G* [151] similar local minimum was not detected.[a]

In [152] within the SCF/4-21G approximation was demonstrated that both the cyanate and fulminate anions can be coordinated to Li^+ by both their ends forming the linear structures LiNCO(IV) and NCOLi(VIII), LiCNO(IX) and CNOLi(XIII), see Figure 9. For the cyanate, the coordination by N and for the fulminate by O is more advantageous, the alternative NCOLi and LiCNO lie above them about ~ 10 kcal/mol.

Profiles of PES'es along the MEP of Li^+ and Cu^+ migration around NCO^- and NCS^- are drawn on Figure 10. The optimized geometry and energy characteristics of their special points are given in Table 9.

As it can be seen from Figure 10, a shape of PES'es of the LiNCY → NCYLi and LiCNY → CNYLi (Y = 0 and S) isomerization changes when replacing O by S in a following way. For the rodanide, as for cyanate, the linear coordination IV by N is preferable. The second isomer VII with Li^+ located above the C—S bond, is by ~ 5 kcal/mol higher, the linear VIII structure (Li^+ bonded with S) is a top of the high inversion barrier ~ 18 kcal/mol and not stable to the conversion into VII. IV and VII are separated by small barrier ~ 3–4 kcal/mol. For the cyanate the NCOLi structure is linear.

Contrary to the NCS^-, the CNS^- anion is coordinated to Li^+ by the S—N bond: XI is a ground structure, linear IX with coordinated C atom is 14 kcal/mol higher and separated by a small barrier ~ 5–6 kcal/mol. The linear XIII structure with coordinated S, as VIII for rodanide, is a top of the inversion barrier, contrary to the fulminate anion where the linear CNOLi structure (by O) is ground, coordination by the N—O bond is not typical (there is no special point of PES corresponding to the location of Li^+ above N—O), and linear LiCNO lies ~ 10 kcal/mol higher than CNOLi.

---

[a]A shape of PES in the vicinity of cycle II is very sensitive to computational approximation. For a definite conclusion on the stability of the cyclic intermediates more sophisticated calculations are necessary.

When coming from Li^+ to Cu^+, a qualitative shape of PES'es and structure of their special points remains to be similar, but energies E_{isom} and barriers h_{isom} increases strongly. For CuNCS VII is ~ 15 kcal/mol higher than the ground IV, and the barrier between them is twice more than for LiNCS. For CuNCS IX and XI are close in energy and separated by quite a tangible barrier ~ 17 kcal/mol, contrary to LiCNS, where IX is by ~ 15 kcal/mol less advantageous and separated by a small barrier ~ 5 kcal/mol. As usual, compounds of Cu^+ are more rigid, than Li^+, to the migration of the cation around the anion.

If to replace now Li^+ by H^+, the protonation of the cyanate anion is energetically preferable by N (NCOH is ~ 23 kcal/mol higher than the ground HNCO), but for the fulminate the proton affinity of both ends is very close: HCNO and CNOH are almost quasi-degenerated [153]. The structures with the protonated central atom are very non-favorable (~ 50 kcal/mol higher) and sometimes can be intermediates [152–154]. For the rodanide, NCSH is about ~ 10 kcal/mol higher as compared to the ground HNCS, for CNS^- protonation of C is more profitable than S [150,151].

As it was demonstrated in Section 3.3 for the MAB complexes with diatomic ligands, a solvation of a cation decreases usually isomerization barriers and flattens the sections of PES between isomers. For the isolated LiNCS molecule the barrier V between VI and IV is about 5 kcal/mol [151]. In parallel with the effective charge on Li, an addition of H_2O stabilizes the ground IV isomer stronger than the other, V and VI, by 2 and 3–4 kcal/less. These effects per one H_2O molecule are small but they accumulate with increases of n and can diminish the barrier altogether. Therefore, the isomerism, inherent for the isolated LiNCS molecule, becomes to be much less or not a characteristic for its solvated form $Li^+(H_2O)_n \cdot NCS^-$, in accordance with the experimental study [155] detecting the only coordination by N in a water solution. The isomerism will be even less typical for the compounds of heavy alkali metals with lesser barriers. On the other hand, for complexes of Cu^+ the barriers are essentially larger than for Li^+, and we can suppose the isomerism for solvates of Cu^+ to be more clearly pronounced than for those of the alkali metals (e.g., at the same n, lesser than maximal coordination number of the cation).

3.6.3. <u>Other ligands ABC with 16 valent electrons</u>. There are papers devoted to PES'es of the N_2O [156,157], N_3^- [158,159], P_3^- [???] ligands and LiN_3 salt [160]. Systematical calculations of the valent-isoelectronic N_2S, P_2O, P_2S, NPO, NPS molecules [161] and N_2P^-, NP_2^-, P_3^- anions [162] were carried out as well. Profiles of their PES'es are plotted on Figure 11, alternative structures of their special points are drawn on Figure 12. In Table 10 the optimized values of geometry and energy parameters are collected.

It is seen from Tables 10 and 11, that for the N_2O, N_2S molecules and the N_3^-, N_2P^- anions, having two or all three atoms from the second period, a global minimum corresponds to the linear structure I and the cycle III is much less advantageous (by ~ 50–60 kcal/mol for N_2O and N_3^-) because of the very strong $2p\pi(N)$—$2p\pi(N)$ bond, broken in III. It was shown in [156] that for N$_2$O III is a shallow local minimum with a barrier of about a few kcal/mol; for N$_2$S and N$_2$P$^-$ III is close to the dissociation limit. If we were to replace two N atoms by P, I is strongly destabilized and has a trend to approach III in energy for P$_2$O and P$_2$N$^-$, because the $3p\pi(P)$—$3p\pi(P)$ is much weaker when compared with $2p\pi(N)$—$2p\pi(N)$. For the last two systems the linear and cyclic isomers I and III are separated by quite an appreciable barrier II and, probably, can be detected, e.g., in matrix isolation experiments. For P$_2$S and P$_3^-$ I and III are close in energy (for P$_2$S the cycle III is a few kcal/mol lower and for P$_3^-$ slightly higher than the linear I). In accordance with the model [111–113], the cyclic isomer will increasingly predominate in the series As$_2$O—Sb$_2$O—BiO, P$_3^-$—As$_3^-$—Sb$_3^{3-}$—Bi$_3^-$, P$_2$S—As$_2$S$_2$—Sb$_2$S etc. in parallel with the weakening of $np\pi(A)$—$np\pi(A)$ bond.

For these ligands new questions arise: what are PES'es of migration of cation like Li$^+$ and proton around linear I, V and cyclic III isomers of the ABC ligands? What kind of coordinations are advantageous and disadvantageous? How does relative stability of the isomers change from the isolated ligand ABC to the coordinated one in the inner sphere of the cation? When the cycles are stabilized or destabilized and have a trend of opening when coordination occurs? etc. These questions are discussed in detail in [163,164].

3.7. Complexes of Triatomic ABC Ligands with 18 and 20 Valent Electrons

3.7.1. <u>Anions NO$_2^-$, NS$_2^-$, PO$_2^-$ and PS$_2^-$, molecules O$_3$,SO$_2$, S$_2$O, S$_3$ and their analogs.</u> Experimental investigations of oxianions NO$_2^-$, PO$_2^-$, PsO$_2^-$, SbO$_2^-$ and molecules O$_3$, SO$_2$ SeO$_2$, TeO$_2$ justify their bent structure I (see Figure 13), with $\varphi(OA)$ varying within 100–120° [48]. Nonempirical calculations of PES for O$_3$ [165–168], SO$_2$ [169], S$_2$O [170], S$_3$ [165,166,168], Se$_3$ and Te$_3$ [168] demonstrate a possibility of the cyclic II isomer. They have different electronic configurations, the barrier III dividing I from II is high enough, and in the vicinity of III a crossing of HOMO and LOMO takes place.

Similar to ligands with N_{val} = 16 (see Section 3.5), for ligands with N_{val} = 18, having two or all three atoms from the second period, the cyclic II isomer is much less favorable (by ~ 25–35 kcal/mol for O$_3$, 90 kcal/mol for SO$_2$, etc.) than the ground I. When all three light atoms are replaced by the heavier analogs, energies of π-bonds sharply lessen, I is destabilized relative to II, resulting to approaching I and II in energy. For S$_3$ II is only a few kcal/mol less advantageous than I, for Se$_3$ and Te$_3$ I and II are quasi-degenerated, similar to the series P$_2$O—As$_2$O—Sb$_2$O discussed in Section 3.5.3. For S$_2$O cycle II becomes more stable by ~ 8 kcal/mol than I with central O atom, but the ground structure here is IV with central "four valent" S atom, having S=O and S=S bonds.

What are coordinations of this ligands to cations like Li$^+$?

An experimental study of the isolated molecules of alkali metal salts MNO_2, MPO_2, etc. [48,62] demonstrate their symmetrical cyclic structure C_{2v} (structure V on Figure 13) with bidentate coordination of cation to both O atoms. Non-empirical calculations of $LiNO_2$ [171,172] approve V to be a global minimum of PES and IX (Li^+ is bonded with N by means of a lone electron pair of the latter)—a top of the potential barrier about 30 kcal/mol.

For an electrostatic reason, this barrier will grown sharply from MNO_2 to MOP_2 and then more slowly in the series MPO_2—$MAsO_2$—$MSbO_2$—$MBiO_2$. In isolated molecules of the alkali metal salts IX scarcely can exist. PES of $LiNO_2$ has a shallow local minimum VII with Li^+ coordinated from outside to N—O bond in the anion's plane, but the barrier VI separating VII from V, is only ~ 1–2 kcal/mol, and more sophisticated calculations are necessary for definite conclusion on the ability of VI to exist. The cyclic V structure is non-rigid to a large amplitude movements of the cation in its vicinity in plane and perpendicular to the plane of the anion AO_2^-.

For ion-molecular complexes of Li^+ with O_3, SO_2, SeO_2 and TeO_2 the deformationally non-rigid bidentate structure V remains to be ground and advantageous. The plane IX structure where Li^+ is coordinated to the central A atom which positive effects increase from SO_2 to TeO_2, is not stable to the conversion in to V or to the dissociation $Li^+ + AO_2$.

When passing from dioxides AO_2 to disulphides AS_2 and the more to diselenides ASe_2, the decrease of the positive charge on A will be accompanied by the stabilization of IX and the destabilization of VI (with the linear or bent A—S_b—M fragment) as compared to the bidentate V. For $LiNS_2$ additional isomers with the $N{\equiv}S$—S^- skeleton appear, IX can be close to V or even more profitable. For complexes with S_3, Se_3 and Te_3 several new low-lying structures with the cyclic ligand (Li^+ is coordinated to the vertex, the edge or face of the triangle as η^1, η^2 and η^3, correspondingly) are possible as well.

What is the situation for HAO_2, HAS_2, etc. acid molecules?

In accordance with experiment [48] and calculations [173,174], HNO_2 has a ground *cis*-structure VIII HONO (its *trans*-form VII is only 1 kcal/mol higher), the excited IX (H—NO_2) is less advantageous and separated by a high barrier. Replacing the Li^+ with a proton results in the stabilization of IX and the destabilization of V, the latter becomes to be a top of the barriers. For HPO_2 the qualitative picture is similar, but, because of the P—H bond, energy is ~ 15 kcal/mol less than N—H, IX appears to be ~ 20 kcal/mol less favorable than *cis*-form HOPO (VIII); its *trans*-form VII lies by 4 kcal/mol higher than *cis*.

The S—H bond energy is ~ 25 kcal/mol less as compared with O—H. From HNO_2 to HNS_2 structures IX and VIII are inverted on the energy scale, and IX comes to be ~ 10 kcal/mol more advantageous. *Cis*-structure HN=S=S(X) is very close to IX with 1 kcal/mol. The XI cycle is close to *cis*-VII and only 10 kcal/mol higher than the quasi-degenerated ground IX and X. It can be seen that substitution of both O atoms by S or heavier

chalkogens is accompanied by essential destabilization of the "classical" VIII structure with S-H bond and the stabilization of structures within the N-H bond, H with N=S=S skeleton and XI cycle. The KNS_2 molecule has several close-lying (within 10-12 kcal/mol) isomers, separated by barriers and very different by their geometric and electronic structures. They are related to the S=N$^-$=S $\rightleftarrows$ N$^-$=S=S isomerism in the anion, formation of
the N$^-$ < $\begin{smallmatrix} S \\ | \\ S \end{smallmatrix}$ cycle and protonation of the bent NS_2^- by N and S atoms. Many of them can probably be detected experimentally. The picture will be even more complicated if you take into consideration low-lying triplet states (here we limited ourselves with the lowest singlet only).

The HNSO molecule [175,176] has a ground *cis* -structure HN=S=O(X) with tetra-valent S in central position. Protonation of NSO$^-$ by the O atom with the forming of N$\equiv$S—OH is less advantageous by 10 kcal/mol. The HS—N=O structures lies ~ 10 kcal/mol higher than X as well. Protonation of SNO$^-$ by O is by 11 kcal/mol less profitable than by S. Resuming, the substitution of one O atom by S, when passing from HNO_2 to HNOS is accompanied by the appearance of several close-lying isomers for sulphur compounds, separated by barriers. The isomers here related to the O=N$^-$=S $\rightleftarrows$ N$^-$=S=O isomerism in anion and protonation of each of them by both ends. One can assume them to be detected experimentally as well.

That is probably a peculiarity of the nitrogen chalkogenides, where isomers like N$\equiv$S—S$^-$ and N$\equiv$S—O$^-$ with multiple N—chalkogen bond are specially favorable. If we were to change N by P or As, similar structures like P$\equiv$S—S$^-$ or As$\equiv$S—S$^-$ are strongly destabilized while XI is stabilized. Therefore, for chalogenides of P, As and heavier atoms of the Vth group the *cis* - and *trans* - forms VII, VIII and cycle XI will be compatible.

Surely, full-scale non-empirical calculations and experiments are necessary to justify these predictions.

For salts of halogen dioxides like $MClO_2$, $MBrO_2$ and MIO_2 the bidentate V structure is most probable, while for their acids $HClO_2$, $HBrO_2$ and HJO_2 V will be less advantageous as compared with VII or VIII.

4.1. IONS M_2AB^+ WITH TWO CATIONS

Ions M_2AB^+ are of interest in two aspects. On one side, intensive peaks of ions like Li_2CN^+, Li_2BOP^+, $LiAlO^+$ in the isolated state are observed in mass-spectra of high-temperature vapor above the melted LiCN, LiBO etc. salts whose molecules have high enough (~ 40–50 kcal/mol) affinity to alkali metal cations [48,51,177]. It is important to know their structure, vibration frequencies and disposition to rearrangements and migration of two cations around the ligand. On the other hand, these systems can serve as models of fragments in complex compounds, with "bridged" ligand AB bonded with two cations. The ligand can coordinate to both of them either by one of its ends (see Figure 14, structures I and VIII), or by different ends to different cations

(III and IV), or by one atom to one cation and by bond to another one (II and VII), or by bond to both cations (plane V and "butterfly"-like VI). It is interesting to know relative energies of structures I—VIII, their geometric and electronic structure and barriers between them for various combinations of M^+ and AB, which can be determined theoretically by scanning of PES in the vicinity of I—VIII and between them. As before A is electropositive and B electronegative atoms of AB ligand.

Non-empirical calculations are known for the Li_2BO^+, Li_2AlO^+, Li_2BS^+ and Li_2AlS^+ ions [178], and Li_2C_2 [143], Li_2O_2, Na_2O_2, Li_2S_2, Na_2S_2 [179], Li_2F_2 [82] molecules etc. Their results are given in Table 11. We will start with ligands AB containing 10 valent electrons again.

As it was said in Section 3.1, PES of salt molecules LiAB with light anions like BO⁻ and CN⁻ has two minima close in energy, relating to the linear LiAB or ABLi isomers. Within the electrostatic model it is naturally to expect the second cation to join the free ligand's atom forming the linear III structure where the anion is stretched out between the cations and coordinated to them by its different ends. Really, one sees from Table 11, that III is a ground structure of $LiCNLi^+$ and $LiBOLi^+$ the nearest excited isomer VIII (the ligand is coordinated to both cations by its electronegative end) is ~ 20 kcal/mol above for $BOLi_2^+$ and separated from III by low barrier ~ 2 kcal/mol (the structure VIII). For $CNLi_2^+$ III and VIII have to be still closer in energy and the barrier lesser than for $BOLi_2^+$, so that $BOLi_2^+$ and $CNLi_2^+$ are scarcely stable to the conversion into III. The structure I (both cations are bonded with the electropositive and A) is much less advantageous, for Li_2BO^+ corresponds to a top of barrier and is not stable to the conversion into III. The structures II and VII where the ligand is coordinated by bond to one (and the more to both) cation are not stable as well corresponding either to a top of a barrier or to non-stationary points of PES.

Let us underline a difference between the triatomic salts $LiA^{II}B^{II}$ and their $Li_2A^{II}B^{II+}$ ions. While in the formers the coordinations LiAB and ABLi can be close in energy, in the latters the coordination VIII to both cations by the electronegative B end is always much more (15—30 kcal/mol) profitable than I (by the electropositive A end), and the energy difference between I and VIII increases with the growing of the electronegativity difference between A and B, in accordance with the electrostatic model.

The replacement of A^{II} by A^{III} leads to the increase of the negative effective charge on the B atom and to sharp destabilization of the structures like LiAlO and LiSiN; their alternative AlOLi and SiNLi are ground and dominating. Naturally, this tendency is preserved for the Li_2AB^+ ions: the "anion" VIII structures $AlOLi_2^+$ and $SiNLi_2^+$ with the central electronegative atom O or N, appear to be ground and the linear III—excited. For $AlOLi_2^+$ III is ~ 13 kcal/mol above VIII, separated by the quite appreciable barrier ~ 15 kcal/mol and will be stable up to a moderately high temperature. The structure I (Li_2AlO^+ and Li_2SiN^+) is even less stable. On the electrostatic ground, energetic stability of VIII against III will increase (if the B atom is fixed) when A is changing in the series Al—Ga—In—Te and Si—Ge—Sn—Pb. Surely, the VIII structure $AOLi_2^+$, $ASLi_2^+$ will be most advantageous, if A is transition metal.

Contrariwise, replacing B^{II} by B^{III} (O by S) stabilizes the linear LiBS and LiCP structures for salts and I for the Li_2BS^+ and Li_2CP^+ ions, which approaches VIII in energy or even becomes more advantageous than VIII. The ground isomer of Li_2BS^+ is IV where the $\varphi(Li_{(1)}BS)$ angle is close to 180° and $\varphi(Li_{(2)}SB) \sim 120°$. IV lies by 17 and 22 kcal/mol below than I and VIII, correspondingly. The barriers between I, IV and VIII are very low. For Li_2CP^+ I and IV are closer and VIII is less stable when compared with Li_2BS^+. The structures I and IV will approach in energy (up to inversion) when decreasing of electronegativity in the series like S—Se—Te and P—As—Sb—Bi.

At last substitution of both light atoms $A^{II} \rightarrow A^{III}$ and $B^{II} \rightarrow B^{III}$ is accompanied by the approach of the IV and VIII structures in energy, and the destabilization of I for $AlSLi_2^+$ and $SiPLi_2^+$. VIII is a ground isomer, lying by ~ 10 kcal/mol below than III.

It is worth mentioning that PES'es of $LiBSLi^+$ and $LiAlSLi^+$ are very flat between the III and IV structures. III is a top of small inversion barrier ~ 1–2 kcal/mol, and $Li_{(2)}^+$ can perform the large-amplitude movements around the S atom and partially above the A—S bond. The inversion barrier will increase in the series S—Se—Te.

Let us compare these results again with the data for the relative HAB molecules and H_2AB^+ ions calculated in [143,181–185]. As it was told in Section 3.2, HCN has two linear HCN and CNH isomers. For the H_2CN^+ ion [143,181,182] the linear $HCNH^+$ structure III is most advantageous, VIII and I are above it by 40 and 70 kcal/mol. As for Li_2AB^+, protonation of the $HA^{II}B^{II}$ molecule results in the stabilization of the VIII structure having both atoms H, bonded with the electronegative B atom, more stronger as compared with I.

When changing N by P the structure I with two H at the C atom becomes ground, the linear $HCPH^+$ (III) is 45 kcal/mol above and, as for CPH, corresponds to a top of the isomerization barrier [183] and not stable to the conversion into I. If it were to further replace P by As, Sb and Bi, I will be more and more preferable. When substituting C by Si, contrarily, the structure VIII with two H at the electronegative atom (e.g., $SiNH_2^+$) can be ground one. Using the same tendencies one can assume IV to be the ground structure for $HBOH^+$, $HBSH^+$, $HAlOH^+$ and $HAlSH^+$.

It was shown in Sections 3.4 and 3.5, that the LiAB molecules with 12 and 14 valent electrons in singlet state have a cyclic ground structure with Li^+ located above the A—B bond. From an electrostatic point of view, the Li_2NO^+, Li_2NS^+, Li_2PO^+, Li_2PS^+ ions (in the singlet state), on one side, and Li_2OF^+, Li_2OCL^+, Li_2SF^+ and Li_2SCl^+ on another will have the bicyclic ground structure ("butterfly"-like VI for $A^{III}B^{III}$ and plane for $A^{II}B^{II}$ and A^{II}) with the ligand coordinated by the bond to both cations. This prediction is in accordance with data of Table 11B for the ions Li_2AB^+ with $N_{val} = 14$ and results for the Li_2O_2, Na_2O_2, Li_2S_2, Li_2S_2, Na_2S_2 [190], Li_2C_2 [25,143,180], and related Li_2A_2 molecules calculated nonempirically.

On PES of Li_2C_2 there are two minima, a global plane bicyclic V (D_{2h}) and a local linear III ($D_{\infty h}$) lying ~ 10 kcal/mol above. The barrier is low and Li_2C_2 can be regarded as a non-rigid molecule toward large shifts of both cations along the C—C bond. A consequent movement of the cations is energetically more preferable as compared with a synchronous one. The structure I(C_{2v}) with both Li^+ at the same C atom *trans*-forms when geometry optimization into V. A global minimum for Li_2Si_2 (singlet) corresponds to the "butterfly" structure VI with the dihedral angle ~ 100° which is a non-rigid coordinate: the inversion barrier via the plane V structure does not exceed a few kcal. The linear III structure ($D_{\infty h}$) is less stable by ~ 30 kcal/mol.

Li_2O_2 has the plane V and Li_2S_2 the folded VI bicyclic structure. For Li_2S_2 VI has the dihedral angle ~ 120° and, as for Li_2Si_2, is non-rigid toward inversion through V being a top of a low barrier. The linear I structure of Li_2O_2 and Li_2S_2 is much more disadvantageous.

Both $Li_2A_2^{II}$ and $Li_2A_2^{III}$ prefer the bicyclic structure with the coordination of the A_2^{2-} anion "by bond" to both cations, differing from their hydrogen analogs $H_2A_2^{II}$ and $H_2A_2^{IOI}$ with terminal H—A bonds and direct A—A bond, single or multiple. The bicyclic structure predominates for the electropositive cations like Liu^+. When increasing electronegativity of M^+ (e.g., from Na^+ to MgH^+, AlH_2^+ and so on) classical structures like III or IV are stabilized stronger, than V and VI, and can be ground as for the H_2A_2 hydrides.

A difference between $Li_2A_2^{II}$ and $Li_2A_2^{III}$ is that the bicyclic structure is planar (V) for the formers and folded (VI) and non-rigid toward the inversion (opening of the "butterfly") via V for the latters. The folded structure VI probably will be more stable for the lithiated compounds of more heavier atoms (A^{IV}, A^V) like Li_2Ge_2, Li_2Sn_2, Li_2Se_2, etc. in parallel with the increase of the Li—Li distance and covalency of interaction between $2p\pi(Li)$—AO and $\pi(A_2)$—MO.

One sees from Table 11B, that the Li_2AB^+ ions with $N_{val} = 14$ have the bicyclic structures as well (V for Li_2OF^+ and Li_2SF^+, VI for Li_2SCl^+); I and VIII are scarcely stable toward transformation into V. An exception is Li_2OCL^+ having the ground I structure with both Li^+ coordinated to O, in accordance with deformational non-rigid structure LiOCl (see Section 3.6).

Dimers of alkali metal halogenides like Li_2F_2 and Li_2Cl_2 have the planar rombic structure IX (D_{2h}) with an acute angle at the halogen atom [48,82].

4.2 IONS M_2ABC^+

As it was demonstrated in Section 3.6, the salt neutral molecules with light triatomic anions like BO_2^-, NCO^-, CNO^-, NCN^{2-} and N_3^- have two linear isomers M[ABC] and [ABC]M with the ligand, coordinated to the cation by one of its terminal atoms A or C. Other structures with the cation, located above the central B atom or above A—B or B—C B—C bonds, correspond either to a top of barrier or to non-stationary points of PES. Therefore, because of an electrostatic reason, for the ions $LiOBLi^+$, $LiNCOLi^+$, $LiCNOLi^+$, $LiNCNLi$, $LiNNNLi^+$ etc. one can wait for the linear ground structure V (see Figure 15) with the ligand coordinated by its different ends to different cations. The next excited structure has to be XIII $OBOLi_2^+$, $NCOLi_2^+$, $CNOLi_2^+$, etc., with both Li^+ coordinated to the electronegative end O (or N), while the structure I (Li_2NCO^+ and Li_2CNO^+, both cations are bonded with the electropositive end) lie higher and are much less stable.

PES of the LiOBS and LiOAlS molecules has a single minimum (linear structure I on Figure 7), therefore for the Li_2OBS^+ and Li_2OAlS the structure XII (Figure 15) with both Li^+, coordinated with the electronegative O, will be ground and more and more advantageous in the series B—Al—Ge—In. For the $LiBS_2$ molecule a single minimum of PES is related with the cyclic structure III on Figure 7, and for the $Li_2BS_2^+$ ion the structure III (Figure 15) is probable, against VI and XII for $Li_2AlO_2^+$.

The LiNCS molecule has, above the linear ground isomer, a shallow local minimum (structure XI on Figure 9 with Li^+ coordinated to the C—S bond) lying above by ~ 5 kcal/mol only. For the Li_2NCS^+ ion a probable ground structure is XII (Figure 15), while X and III with one or both Li^+ above the C—S bond are low-lying. For Li_2CNS^+ III and XI (one Li^+ is bonded with C and second Li^+ above the N—S bond) are compatible.

Again a question arises, in what degree the conclusions, made above for isolated ions and salt molecules, can be transferable onto "real" complexes containing the same MABM or MABCM fragments in their coordination sphere. Geometry of isolated species was fully optimized for every alternative configuration. Therefore, the equilibrium R(MM) distance between cations, depended on a balance between their attraction to the anion and repulsion from each other, can vary strongly from structure to structure. For example, R(LiLi) in the I, III and VI of the Li_2BO^+ ion (see Figure 14) is ~ 3.5, 5.4 and 3.7 Å, correspondingly; for realization of the ground linear structure III it is necessary to move the Li^+ cations apart by ~ 5 Å or more.

In "real" coordination compounds the cation—cation distance can be much shorter, especially if the cations are bonded to each other by additional interactions (by means of other bridged ligands, by direct metal—metal bonds; for crystal or for an active center on a surface the steric effects of lattice or base are important). An optimal coordination of anion to two cations has to adjust to these short R(MM) distances. Therefore the most advantageous coordination and relative position of other alternative structures on the energy scale can differ from those for isolated fragments MABM and MABCM, especially if the structures under discussion are close in energy for isolated fragments. Results depend on R(MM) and limits of its variations in certain complexes. Nevertheless it seems to be reasonable to suppose that the coordinations, which are obviously disadvantageous for isolated fragments, will remain disadvantageous in complexes. The most pronounced tendencies of relative energies of alternative coordinations (e.g., when varying the ligand in the series like CN⁻—CP⁻—CAs⁻—SiN—SiP...at invariable cations M) will be preserved as well. The best approximate transferability can be waited for the complexes with a single bridged ligand between cations and without additional bonding interactions metal—metal. One can hope that the chemical environment of the cations, if their coordination sphere is filled with neutral ligands like CO, H_2O or NH_3, will not violate a qualitative picture as we have observed for the fragments with one cation in Section 3.3.

4.3 DIMERS $(MAB)_2$

Isolated dimer molecules were detected by a mass-spectrometry method in a high-temperature vapor above many melted salts, but detailed structural data is available mainly for the alkali metal halides [72]. Information for molecules like $(MAB)_2$ and $(MABC)_2$ with diatomic and triatomic ligands is scarce. IR spectra of the matrix isolated lithium cyanide dimer were interpreted in [65,66] in favor of the structure Ia(D_{2h}) on Figure 16 with the four-member $[N_2Li_2]$ cycle and two "terminal" C atoms, while for $(NaCN)_2$ and $(KCN)_2$-in favor of II (D_{2h}) with the $[C_2M_2]$ cycles and the terminal N atoms. The SCF study of the $(LiNC)_2$ dimer [186], with using the 4-31G basis for geometry optimization and 6-31G* for the single-point calculations of relative energies, demonstrated the ground structure to be Ia, in accordance with experiment [65,66], the six-member cycles IV and V are very close in energy and lie ~ 4 kcal/mol above Ia; III is ~ 22 kcal/mol higher than Ia. The later SCF/DZHD + P calculations [187] pointed out that the ground structure I is deformational non-rigid: deviations of the φ(CNN) angle from 180° by 30—40° are accompanied by very small changing of the total energy by a few tenths of kcal/mol. On the SCF level, the symmetric (D_{2h}) structure Ia corresponds to a top of slightly appreciable barrier about 0.5 kcal/mol between two shallow minima with cis - and trans -positions of the CN groups. From our point of view, this energy difference is too small for quantitative evaluation of the equilibrium φ(CNN) value and of the height of the barrier (for this goal more sophisticated calculations are necessary), and as usual in a situation of this kind, we limit ourselves by the statement of the deformation non-rigidity of dimers like $(LiCN)_2$.

In [188] similar SCF calculations with the 3-21G* and DZHD + P basis sets were performed for the $(LiOB)_2$, $(LiBS)_2$, $(LiOAl)_2$ and $(LiSAL)_2$ dimers. Their results are given in Table 12. One sees that, as for $(CNLi)_2$, $(LiOB)_2$ has the same group of low-lying structures: II with the $[O_2Li_2]$ four-member cycle and terminal boron atoms, and the six-member cycles IV and V quasi-degenerated in energy (orientation of two BO⁻ anions to each other, parallel or antiparallel, does not influence practically on relative stability of IV and V). The structure I with the $[B^2Li^2]$ cycle lies by ~ 35 kcal above and VII with the $[B_2O_2]$ cycle is not stable toward dissociation onto monomers. The $(CNLi)_2$ dimer differs from $(BOLi)_2$, that for the former, the ground isomer is II lying 5 kcal/mol below IV and V, while for the latter, contrarily, the six-member cycles are ground with II lying above them by 4 kcal/mol. Again, we meet the tendency: coordination of anions like CN⁻ and BO⁻ by both ends to one cation makes the LiAB and ABLi isomers to be close in energy, while their coordination to two cations by electronegative atom B is much more profitable than by the electropositive A end. The relative energy increases from the Li_2BA^+ ions to the $(ABLi)_2$ dimer and with the increase difference of electronegativity between A and B.

In the vicinity of the II isomer PES of the $(BOLi)_2$ dimer is very flat to the deformation of the $\varphi(BOO)$ angle from 180° to 130°. It is not surprising: in Section 3.1 we underline that PES'es of the BOLi and CNLi monomeric isomers are also very flat (when changing $\varphi(BOLi)$ and $\varphi(CNLi)$ from 180° to 130° their energies decrease within a few tenths of kcal/mol only) and sensitive to computation approximation.

If to replace B by Al, the structures I (with the $[Al_2Li_2]$ cycle), IV and V of the $(AlOLi)_2$ dimer are strongly destabilized, and VII (with the $[Al_2O_2]$ cycle) is stabilized as compared with I containing the $[O_2Li_2]$ cycle and appearing to be ground. VII is higher by ~ kcal/mol. I and VII will be most advantageous for $(LiOGa)_2$, $(LiOIn)_2$ and $(LiOTe)_2$.

When substituting O by S the structure II (the cycle $[S_2Li_2]$) is destabilized stronger than IV, V and I ($[B_2Li_2]$). The six-member cycles are ground for $(LiBS)_2$, I and II lie by ~ 10 and ~ 40 kcal/mol above. The structure VII with the $[B_2S_2]$ cycle is not stable. In the series $(LiBS)_2—(LiBSe)_2—(LiBTe)_2$, IV and V are ground isomers, II will be more and I less stable.

For $(LiSAl)_2$ the six-member cycles lie by ~ 25 kcal/mol above the ground structure II ($[Li_2S_2]$), I is not stable to dissociation.

Let us compare the data of Table 12 with the results of similar SCF/3-21G* and MP3/DZHD + P calculations of the $(HBO)_2$, $(HAlO)_2$, $(HBS)_2$ and $(HAlS)_2$ dimers [189]. If we were to change Li⁺ by H⁺, we have quite a different picture. In all cases under discussion, the most advantageous appears to be the structure $(HA)_2B_2$ (VI) with the four-member cycle $[A_2B_2]$ and the H atoms bonded with the electropositive A atoms.[a] When dimerization occurs, the $[A_2B_2]$ cycles are built because of the breaking of two π- and the formation of two δ-bonds A—B. Dimerization energy depends mainly on the difference of energies of these δ(A—B) and π(A—B) bonds and varies from 30 to 70 kcal/mol from $(HBO)_2$ to $(HAlS)_2$. The structures containing bridged hydrogen bonds with both four-member I—III and six-member IV—V

---

[a]Dimers $(H_2Cs)_2$, $(H_2SiO)_2$ and $(H_2SiS)_2$ have similar $(H_2A)_2B_2$ ground structure, in accordance with calculation s[185].

cycles are weakly bonded. The structure $A_2(BH)_2$ (VII) with the $[A_2B_2]$ cycle and H atoms bonded to the electronegative O and S atoms, is not stable for the boron compounds, close to dissociation limit for $(LiSAl)_2$ and correspond stable excited isomers only for $(LiOAl)_2$ lying ~ 25 kcal/mol above VI. In the series $(GaOH)_2$—$(InOH)_2$—$(TlOH)_2$ VII will be stabilized more than VI, and for $(TlOH)_2$ the hydroxide VII structure can be ground one and close to VI.

These differences between H and Li compounds probably are preserved for trimers, tetramers and more complicated aggregates. For example, one can assume the $(HBO)_3$ and $(HBS)_3$ trimers to have a structure of 6-member rings $[B_3O_3]$ and $[B_3S_3]$ where H is bonded with boron atoms. For $(BOLi)_3$ the 9-member rings are strongly destabilized when compared with $[O_3Li_3]$ etc. The 6-member cyclic structure $[O_3Li_3]$ and $[O_3A_3]$ will approach each other in the energy of the series A=Al—Ga—In—Tl.

For dimers $(MABC)_2$ with triatomic ligand a picture is naturally more complicated. For example, the possible candidates to preferable structures for $(LiOBO)_2$ or $(LiOAlO)_2$ can be a) $[Li_2O_2](BO)_2$ with the four-member $[Li_2O_2]$ cycle and two BO groups bonded to the bridged O atoms of the cycle, b) the 8-member cycle, and c) a combination of a) and b)

a)   b)   c)

Above importance for high-temperature chemistry, some dimer molecules like $(MAB)_2$ and $(MABC)_2$ can be approximately regarded as models of fragments in "real" complexes with two bridged ligands between two metal cations.

5. SUMMARY

This review is limited by the simplest fragments cation—ligand and does not pretend to discuss numerous "chemical aspects" of the bond isomerization problem in detail. It is only the first step of the theoretical study of the problem using a language of potential energy surfaces. Its goals were the comparison of structural and energetic characteristics of families of alternative structures and isomers; the finding of ground and low-lying isomers; analysis of tendencies in their relative energies and barriers in the series of similar compounds with varying cations and ligands. Next steps (above the study of the fragments with new ambivalent ligands and cations of transition metals with various oxidation numbers) are 1) expansion of this approach on "real" complex compounds containing these fragments in the coordination sphere together with other ligands; 2) the passing from the "statical" language of potential energy surfaces to the dynamic study of the isomerization; 3) investigation of the role of the environment. The dynamic approach is very essential, because for many fragments cation—ligand, both isolated and in "real" complexes, PES has very flat sections within wide intervals varying from one or several non-rigid geometric coordinates. Structural non-rigid geometric coordinates. Structural non-rigidity plays an important role for understanding the mechanisms of the isomerization. From our point of view, non-rigid systems represent one of the most interesting and promising regions of modern structural chemistry, but at the same time, one of the more difficult both for theoretical and for experimental study. Its understanding can be reached only in a complex approach using of different experimental structural methods with a different characteristic time of measurement and non-empirical calculations of PES, simultaneously. It is true both for isolated molecules as well as for ion-molecular complexes studied in the gaseous phase or in inert matrix isolation by electron diffraction, mass-spectrometric or spectroscopic methods, and for coordination compounds studied in solids or in the solution with the use of x-ray diffraction, IR, NMR and calorimetry methods. Structural and energetic information, now accumulated in these two so far distant areas, actually is related to different aspects of the same problem of the bonding isomerism and complement each other. One of the goals of theory is to create bridges between these areas.

For the isolated MAB, M_2AB^+, $(MAB)_2$ and similar fragments a qualitative electrostatic model was demonstrated to be effective and helpful. In some cases, when a polar nature of the cation—ligand bonds is not essentially changed when the saturation of the coordination sphere by other ligand, the electrostatic approach can be valid for more complex compounds as well. For example, within its limit one can understand, why the structural information available now for complexes with the light ambivalent ligands, consisting mainly from atoms of the second period, can not be transferable onto relative ligands, where two or more light atoms are replaced by their heavier analogs in the same subgroups. When replacement of this kind occurs, different sections of the potential energy surfaces are stabilized or destabilized very strongly, the alternative structures can change their relative position on the energy scale up to the inversion and change of the structural type of the ground isomer; new isomers can arise. Coordinations profitable for the light ligands often become non-advantageous for the heavier ligands and correspond to nonstationary points of PES or to the top of potential barrier.

Together with non-empirical calculations of PES, the electrostatic model can be useful to figure out where the periodicity rule is violated, where it is preserved and why, and to predict structures of compounds of elements from the fourth, fifth and heavier periods, not yet studied.

We believe that the non-empirical calculations of the potential energy surfaces will stimulate experimental investigations of the bonding isomerism, not only explaining the known facts but suggesting predictions in cases where the experiment is difficult, non-complete or altogether absent.

REFERENCES

1. S.M. Jorgensen, *Z. Anorg. Allg. Chem.*, 1893, B. 5, S. 169.
2. Yu.N. Kuckuschkin, *Koordin. Khimija* (Russ.), 1978, Vol. 4, No. 8, p. 1170.
3. Yu.N. Kuckuschkin, *Zh. Neorgan. Khimii* (Russ.), 1984, Vol. 29, No. 1, p. 3.
4. M.A. Porai-Koshitz, *Tzintzandze G.V. INT, Ser., "Crystallography 1965"* (Russ.), VINITI, Moscow, 1967, p. 168.
5. I. Lewis, R.S. Nyholm and P.W. Smith, *J. Chem. Soc.*, 1961, p. 4590.
6. A. Tramer, *J. Chim. Phys.*, 1962, Vol. 52, p. 232.
7. P.C.H. Mitchell and R.J.P. Williams, *J. Chem. Soc.*, 1960, p. 1912.
8. V.P. Tarasov, T.Sh. Kapanadze, G.V. Tzintzandze and Yu.A. Buslaev, *Koordin Khimija* (Russ.), 1983, Vol. 9, No. 5, p. 647.
9. V.P. Tarasov, T.Sh. Kapanadze, G.V. Tzintzadze, Yu.A. Buslaev et al., *Koordin. Khimija* (Russ.), 1984, Vol. 10, No. 3, p. 368.
10. V.P. Tarasov, T.Sh. Kapanadze, G.V. Tzintzadze, Yu.A. Buslaev et al., *Koordin. Khimija* (Russ.), 1985, Vol. 11, No. 3, p. 655.
11. V. Doron, *Inorg. Nucl. Chem. Lett.*, 1968, Vol. 4, p. 601.
12. J.E. Hause and R.K. Bunting, *Thermochim. Acta*, 1975, Vol. 11, p. 357.
13. D.A. Buckingham, I.I. Creaser and A.M. Sargeson, *Inorg. Chem.*, 1970, Vol. 9, p. 655.
14. S. Huzinaga, *J. Chem. Phys.*, 1965, Vol. 42, p. 1293.
15. T.H. Dunning, *J. Chem. Phys.*, 1970, Vol. 53, p. 2823.
16. B. Roos, A. Veillard and G. Vinot, *Theoret. Chim. Acta*, 1971, Vol. 20, No. 1, p. 1.
17. B. Roos and P. Siegbahn, *Theoret. Chim. Acta*, 1970, Vol. 17, No. 2–3, p. 209.
18. A.G.H. Wachters, *J. Chem. Phys.*, 1970, Vol. 52, No. 3, p. 1033.
19. N.M. Klimenko, D.G. Musaev and O.P. Charkin, *Zh. Strukt. Khimii* (Russ.), 1985, Vol. 26, No. 5, p. 3.
20. U. Salzner and P.v.R. Schleyer, *J. Comput. Chem.*, 1990, Vol. ___, No. ___.
21. M.V. Ramana and D.H. Phillips, *J. Chem. Phys.*, 1988, Vol. 88, No. 4, p. 2637.
22. C.J. Nelin, P.C. Bagus and M.R. Philpott, *J. Chem. Phys.*, 1987, Vol. 87, No. 4, p. 2170.

23. M. Merchan, I. Nebot-Gil, R. Gonzales-Lugue and E. Ortle, *J. Chem. Phys.*, 1987, Vol. 87, No. 3, p. 1690.
24. P.-Y. Margantini and J. Weber, *J. Mol. Struct.* (THEOCHIM), 1988, Vol. 166, p. 247.
25. O.P. Charkin, *Stability and Structure of Gaseous Inorganic Molecules, Radicals and Ions* (Russ.), Nauka, Moscow, 1980, 278 pp.
26. O.P. Charkin and A.I. Boldyrev, *Potential Energy Surfaces and Structural Non-Rigidity of Inorganic Molecules.* (Russ.), INT, Ser. "Neorgan. Khimia", Vol. 8, VINITI, Moscow, 1980, 156 pp.
27. N.G. Rambidi, *J. Mol. Struct.*, 1975, Vol. 28, p. 77.
28. V.A. Istomin, N.F. Stepanov and B. Zhilinskii, *J. Mol. Spectrosc.*, 1977, Vol. 67, p. 267.
29. R.M. Benito, F. Borondo, J.-H. Kim, B.G. Sumpter and G.S. Ezra, *Chem. Phys. Lett.*, 1989, Vol. 161, No. 1, p. 60.
30. J. Tennyson, G. Brocke and S.C. Farantos, *Chem. Phys.*, 1986, Vol. 104, p. 399.
31. M.L. Klein, J.D. Goddard and D.G. Bounds, *J. Chem. Phys.*, 1981, Vol. 75, p. 3909.
32. P.E.S. Wormer and J. Tennyson, *J. Chem. Phys.*, 1981, Vol. 75, No. 3, p. 1245.
33. G. Brocks, J. Tennyson and A. Van Der Avoird, *J. Chem. Phys.*, 1984, Vol. 80, p. 3223.
34. J.J. Van Vaals, W.L. Meerts and A. Dynamus, *J. Chem. Phys.*, 1982, Vol. 77, p. 7245.
35. T. Törring, J. P. Bekooy, W.L. Meerts, J. Hoeft et al., *J. Chem. Phys.*, 1980, Vol. 73, p. 4875.
36. E. Van Leuken, G. Brocks and P.E.S. Wormer, *Chem. Phys.*, 1986, Vol. 110, No. 2,3, p. 365.
37. G. Brocks, *Chem. Phys.*, 1987, Vol. 116, No. 1, p. 33.
38. L.T. Redmon, G.D. Purvis and R.J. Bartlett, *J. Chem. Phys.*, 1980, Vol. 72, p. 986.
39. J.J. Van Vaals, W.L. Meerts and A. Dynamus, *Chem. Phys.*, 1983, Vol. 82, No. 3, p. 385.
40. K. Nakamoto, D. Teroult and S. Tani, *J. Mol. Struct.*, 1978, Vol. 43, p. 75.
41. K. Skudlarski and M. Miller, *Int. J. Mass-Spectr. Ion Phys.*, 1980, Vol. 36, No. 1, p. 19.
42. Z.K. Ismail, R.H. Hauge and J.L. Margrave, *J. Chem. Phys.*, 1972, Vol. 57, No. 12, p. 5137.
43. L.S. Kudin, A.V. Gusarov and L.N. Gorochov, *Teplophysika vysokich temperatura* (Russ.), 1974, Vol. 12, No. 3, p. 509.
44. A.W. Castleman, K.I. Peterson, B.L. Upschulte and F.J. Schelling, *Int. J. Mass-Spectr. Ion Phys.*, 1983, Vol. 47, p. 203.
45. S. Kita, H. Tanuma and M. Izama, *Chem. Phys.*, 1988, Vol. 125, No. 2,3, p. 415.

46. K. Eller, D. Sülzle and H. Schwarz, *Chem. Phys. Letters*, Vol. 154, No. 5, p. 433.
47. Yu.S. Nekrasov, Yu.A. Borisov, D.I. Zagorevskii et al., *J. Organomet. Chem.*, 1984, Vol. 269, No. 3, p. 323.
48. "Molecular constants of inorganic compounds" (Russ.), Ed-r Krasnov K.S. Khimija, St. Petersburg, 1979, 448 pp.
49. V.N. Kondratjev, "Dissociation bond energies. Ionization potentials and electron affinity", (Russ.), Nauka, Moscow, 1974, 351 pp.
50. L.V. Gurvich, I.V. Veitz, V.A. Medvedev et al., "Thermodynamic properties of individual compounds." (Russ.), 3rd edition, Nauka, Moscow, 1981.
51. R.D. Levin and S.G. Lias, "Ionisation potentials and appearance potentials measurements.", 1971–81. NSRDS-NBS 11, Wash. USA, 1982.
52. L. Manceron and L. Andrews, *J. Amer. Chem. Soc.*, 1985, Vol. 107, No. 3, p. 563.
53. E.S. Kline, Z.H. Kafari, R.H. Hauge and J.L. Margrave, *J. Amer. Chem. Soc.*, 1987, Vol. 109, No. 8, p. 2402.
54. G.A. Ozin and A.V. Voet, *Acc. Chem. Res.*, 1973, Vol. 6, No. 9, p. 313.
55. G.A. Ozin and J. Garcia-Prieto, *J. Phys. Chem.*, 1988, Vol. 92, No. 2, p. 325.
56. S.B.H. Bach, C.A. Taylor, R.J. Van Zee, M.T. Vala et al., *J. Amer. Chem. Soc.*, 1986, Vol. 108, No. 22, p. 7104.
57. J. Mascetti and M. Tranguille, *J. Phys. Chem.*, 1988, Vol. 92, No. 8, p. 2177.
58. T.C. De Vore, *Inorg. Chem.*, 1976, Vol. 15, p. 1315.
59. A.M. Shapovalov, V.F. Shevel'kov and A.A. Mal'tzev, *Vestnyk Moskovskogo Universiteta. Ser. "Khimia"* (Russ.), 1975, Vol. 16, No. 2, p. 153.
60. R. Teghil, B. Janis and L. Bencivenni, *Inorg. Chim. Acta*, 1984, Vol. 88, No. 2, p. 115.
61. F. Ramondo and L. Bancivenni, *Mol. Phys.*, 1989, Vol. 67, No. 3, p. 707.
62. M. Barbeschi, L. Bencivenni and F. Ramondo, *Chem. Phys.*, 1987, Vol. 112, No. 3, p. 387.
63. Z.H. Kafafi, R.H. Hauge, W.E. Billups and J.L. Margrave, *Inorg. Chem.*, 1984, Vol. 23, No. 2, p. 177.
64. L. Bencivenni, L. A'Aessio and F. Ramondo, *Inorg. Chim. Acta*, 1966, Vol. 121, no. 2, p. 161.
65. Z.K. Ismail, R.H. Hauge and J.L. Margrave, *High Temp. Sci.*, 1981, Vol. 14, No. 3, p. 197.
66. Z.K. Ismail, R.H. Hauge and J.L. Margrave, *J. Mol. Spectrosc.*, 1975, Vol. 54, No. 3, p. 402.
67. L. Andrews, *J. Chem. Phys.*, 1969, Vol. 50, p. 4288.
68. Yu.S. Ezhov and C.M. Tolmachov, *Zh. Strukt. Khimii* (Russ.), 1984, Vol. 25, No. 3, p. 169.

69. S.A. Komarov and Y.S. Ezhov, *Zh. Strukt. Khimii* (Russ.), 1975, Vol. 16, No. 4, p. 662.
70. V.A. Kulikov, V.V. Ugarov and N.G. Rambidi, *Zh. Strukt. Khimii* (Russ.), 1982, Vol. 23, No. 1, p. 182.
71. V.A. Kulikov, V.V. Ugarov and N.G. Rambidi, *Zh. Strukt. Khimii* (Russ.), 1981, Vol. 22, No. 1, p. 183.
72. J.G. Hartley and M. Fink, *J. Chem. Phys.*, 1988, Vol. 89, No. 10, pp. 6053–6058.
73. L.C. O'Brien and P.F. Bernath, *J. Chem. Phys.*, 1988, Vol. 88, No. 4, p. 2117.
74. C.R. Brazier and P.F. Bernath, *J. Chem. Phys.*, 1988, Vol. 88, No. 4, p. 2112.
75. L.C. Ellingboe, A.M.R.D. Bopegedera, C.R. Brazier and P.F. Bernath, *Chem. Phys. lett.*, 1986, Vol. 126, p. 285.
76. B. Back, E. Clementi and R.N. Koztzenborn, *J. Chem. Phys.*, 1970, Vol. 53, p. 764.
77. E. Clementi, H. Kistenmacher and H. Popkie, *J. Chem. Phys.*, 1973, Vol. 58, p. 2640.
78. A.I. Boldyrev, O.P. Charkin, K.V. Bojenko, N.M. Klimenko et al., *Zh. Neorgan. Khimii* (Russ.), 1979, Vol. 24, No. 3, p. 612.
79. A.E. Smoljar, N.P. Zaretzkii, N.M. Klimenko and O.P. Charkin, *Zh. Neorgan. Khimii* (Russ.), 1979, Vol. 24, No. 12, p. 3165.
80. V. Staemmler, *Chem. Phys.*, 1975, Vol. 7, p. 17.
81. V. Staemmler, *Chem. Phys.*, 1976, Vol. 17, p. 187.
82. A.I. Boldyrev, V.G. Solomonik, V.G. Zakhzevskii and O.P. Charkin, *Zh. Neorgan. Khimii* (Russ.), 1980, Vol. 24, No. 9, p. 2307.
83. A.V. Nemuchin, J.E. Almlöf and A. Heiberg, *Theor. Chim. Acta*, 1981, Vol. 59, No. 4, p. 9.
84. A.Yu. Ermilov, A.V. Nemuchin and N.F. Stepanov, *Zh. Phys. Khimii* (Russ.), 1988, Vol. 62, No. 1, p. 212.
85. T.S. Zjubina, A.S. Zjubin, A.A. Gorbik and O.P. Charkin, *Zh. Neorgan. Khimii* (Russ.), 1985, Vol. 30, No. 11, p. 2739.
86. T.S. Zjubina, O.P. Charkin, A.S. Zjubin and V.G. Zakhzevskii, *Zh. Neorgan. Khimii* (Russ.), 1982, Vol. 27, No. 3, p. 588.
87. V.V. Jackobson, D.G. Musaev and O.P. Charkin, *Zh. Strukt. Khimii* (Russ.), 1986, Vol. 27, No. 6, p. 18.
88. V.V. Jackobson, D.G. Musaev and O.P. Charkin, *Zh. Strukt. Khimii* (Russ.), 1988, Vol. 29, No. 4, p. 20.
89. V.V. Jackobson, D.G. Musaev and O.P. Charkin, *Zh. Strukt. Khimii* (Russ.), 1989, Vol. 30, No. 1, p. 19.
90. D.G. Musaev, A.I. Boldyrev, O.P. Charkin and N.M. Klimenko, *Koordin. Khimiya* (Russ.), 1984, Vol. 10, No. 7, p. 938.
91. O.P. Charkin, D.G. Musaev and N.M. Klimenko, *Koordin. Khimiya* (Russ.), 1985, Vol. 11, No. 4, p. 445.

92. D.G. Musaev, V.V. Jackobson, N.M. Klimenko and O.P. Charkin, *Koordin. Khimiya* (Russ.), 1987, Vol. 13, No. 9, p. 1188.
93. D.G. Musaev, A.M. Mebel, O.P. Charkin, R. Cimiraglia and J. Tomasi, *Koordin. Khimiya* (Russ.), 1989, Vol. 15, No. 9, p. 1155.
94. D.G. Musaev, *Koordin. Khimiya* (Russ.), 1988, Vol. 33, No. 12, p. 3207.
95. D.G. Musaev and O.P. Charkin, *Zh. Neorgan. Khimii* (Russ.), 1991, Vol. 36 (in print).
96. N.M. Klimenko, D.G. Musaev, T.S. Zjubina and O.P. Charkin, *Koordin. Khimiya* (Russ.), 1984, Vol. 10, No. 4, p. 505.
97. M.V. Kuz'minski, A.A. Bagatur'jantz and V.B. Kazanskii, *Izvestija Akad. Nauk of Russia, Ser. Khim.* (Russ.), 1986, Vol. 35, No. 2, p. 284.
98. V.G. Bernstein, N.M. Vitkovskaya and F.K. Schmidt, *React. Kinet. Catal. Lett.*, 1986, Vol. 30, No. 2, p. 361.
99. J. Allison, A. Mavridis and J.F. Harrison, *Polyhedron*, 1988, Vol. 7, No. 16–17, p. 1559.
100. K. Broomfield and R.M. Lambert, *Chem. Phys. Lett.*, 1987, Vol. 139, No. 3,4, p. 267.
101. P.E.M. Siegbahn and M.R.A. Blomberg, *Chem. Phys.*, 1984, Vol. 87, No. 2, p. 189.
102. G.H. Jeung, *Proc. NATO Adv. Res. Workshop a. 40th Int. Meet. Soc. Chim. Phys.*, Strasbourg, Sept. 16–20, 1985. Dortrecht, 1986, p. 101.
103. C. Barbier, G. Berthier, A. Daodi and M. Suard, *Theor. Chim. Acta*, 1988, Vol. 73, No. 5–6, p. 419.
104. G. Berthier, A. Daodi and M. Suard, *J. Mol. Struct.* (THEOCHIM), 1988, Vol. 179, p. 407.
105. C.W. Bauschlicher and L.A. Barnes, *Chem. Phys.*, 1988, Vol. 124, No. 3, p. 383.
106. C.W. Bauschlicher, L.G.M. Petersson and P.E.M. Siegbahn, *J. Chem. Phys.*, 1987, Vol. 87, No. 4, p. 2129.
107. C.W. Bauschlicher, P.S. Bagus, C.J. Nelin and B.O. Roos, *J. Chem. Phys.*, 1986, Vol. 85, No. 1, p. 354.
108. C.W. Bauschlicher, S.R. Langhoff and L.A. Barnes, *Chem. Phys.*, 1989, Vol. 129, No. 3, p. 431.
109. M.R.A. Blomberg, *J. Amer. Chem. Soc.*, 1988, Vol. 110, No. 20, p. 6650.
110. C.E. Dykstra, *Ann. Rev. Phys. Chem.*, 1981, Vol. 32, No. 1, p. 25.
111. O.P. Charkin and T.S. Zjubina, *Koordin. Khimiya* (Russ.), 1986, Vol. 12, No. 8, p. 1011.
112. D.G. Musaev, V.V. Jackobson and O.P. Charkin, *Zh. Strukt. Khimii* (Russ.), 1990, Vol. 31 (to be published).
113. O.P. Charkin, *Zh. Neorgan. Khimii* (Russ.), 1979, Vol. 24, No. 3, p. 581.
114. O.P. Charkin, *Zh. Strukt. Khimii* (Russ.), 1984, Vol. 24, No. 2, p. 87.

115. T.S. Zjubina and O.P. Charkin, *Zh. Neorgan. Khimii* (Russ.), 1984, Vol. 29, No. 3, p. 598.

116. O.P. Charkin, T.S. Zjubina and N.M. Klimenko, *Zh. Neorgan. Khimii* (Russ.), 1984, Vol. 29, No. 7, p. 1635.

117. O.P. Charkin, *Problems of the Theory of Valence, Chemical Bonding and Molecular Structure* (Russ.). Ser. "Khimia", Vol. 7, Zhanie Publ., Moscow, 1987, 47 pp.

118. P.J. Bruno, G. Hirsch, R.T. Beunker and S.D. Peyerimhoff, "Molecular Ions", Ed. Berkowiz., N.Y., Plenum Publ. Corp., 1983, p. 309.

119. F. Kato and K. Morokuma, *Chem. Phys. Lett.*, 1975, Vol. 34, No. 1, p. 7.

120. C.W. Bauschlicher, S.R. Langhoff and H. Partrige, *Chem. Phys. Lett.*, 1985, Vol. 115, No. 2, p. 124.

121. S. Ikuta, *Chem. Phys. Lett.*, 1985, Vol. 95, No. 2, p. 235.

122. S. Ikuta, *J. Mol. Struct.* (THEOCHIM), 1986, Vol. 137, No. 3–4, p. 329.

123. D.G. Musaev, V.V. Jackobson and O.P. Charkin, *Koordin, Khimiya* (Russ.), 1989, Vol. 15, No. 5, p. 588.

124. V.M. Pinchuck, *Zh. Strukt. Khimii* (Russ.), 1985, Vol. 26, No. 3, p. 329.

125. S. Ikuta, *Chem. Phys. Lett.*, 1984, Vol. 109, No. 6, p. 550.

126. D.G. Musaev and O.P. Charkin, *Zh. Neorgan. Khimii* (Russ.), 1991, Vol. 36 (to be published).

127. Th. Weller, W. Meiler, A. Misheal et al., *Chem. Phys.*, 1982, Vol. 72, No. 2, p. 155.

128. Th. Weller, W. Meiler, N.J. Köhler et al., *Chem. Phys. Lett.*, 1983, Vol. 98, No. 6, p. 541.

129. V. Balaji, K.K. Sunil and K.D. Jordan, *Chem. Phys. Lett.*, 1987, Vol. 136, No. 3–4, p. 309.

130. D.G. Musaev and O.P. Charkin, *Koordin Khimiya* (Russ.), 1991, Vol. 17 (to be published).

131. J. Peslak, D.S. Kleh and C.W. David, *J. Amer. Chem. Soc.*, 1971, Vol. 93, p. 5001.

132. V.G. Solomonik and T.P. Pogrebnaja, *Zh. Phis. Khimii*, 1986, Vol. 61, No. 3, p. 1928.

133. G.M. Chaban, N.M. Klimenko and O.P. Charkin, *Zh. Neorgan. Khimii* (Russ.), 1991, Vol. 36 (to be published).

134. N.G. Rambidi, N.F. Stepanov, A.I. Dement'ev and V.V. Ugarov, In "Chemical Bonding and Structure of Molecules" (Russ.). Nauka, Moscow, 1984, p. 5.

135. M.E. Colvin, H.F. Schaefer and J. Bicerano, *J. Chem. Phys.*, 1985, Vol. 83, No. 9, p. 4581.

136. D.G. Musaev and O.P. Charkin, *Zh. Neorgan. Khimii* (Russ.), 1991, Vol. 36 (to be published).

137. J.E. Del Bene, M.J. Frisch, K. Raghavachari, J.A. Pople and P.v.R. Schleyer, *J. Phys. Chem.*, 1983, Vol. 87, No. 1, p. 73.

138. V.V. Jackobson and O.P. Charkin, *Zh. Strukt. Khimii* (Russ.), 1989, Vol. 30, No. 6, p. 19.
139. D.G. Musaev, V.V. Jackobson, A.S. Zjubin, O.P. Charkin et al., *Koordin. Khimiya* (Russ.), 1989, Vol. 15, No. 11, p. 1478.
140. S. Scheiner, *J. Mol. Struct. Theochim.*, 1989, Vol. 200, p. 117.
141. D.G. Musaev, V.V. Jackobson and O.P. Charkin, *Zh. Neorgan. Khimii* (Russ.), 1989, Vol. 34, No. 8, p. 1946.
142. D.G. Musaev, V.V. Jackobson and O.P. Charkin, *Zh. Neorgan. Khimii* (Russ.), 1990, Vol. 35, No. 3, p. 704.
143. R.A. Whiteside, M.J. Frisch and J.A. Pople, *The Carnegie-Mellon Quantum Chemistry Archive. Dept. Chem. Carnegie-Mellon University*, Pittsburgh, Penn., USA, 1983, 183 pp.
144. S.P. Konovalov and V.G. Solomonik, *Zh. Neorgan. Khimii* (Russ.), 1984, Vol. 29, No. 7, p. 1655.
145. M.T. Nguyen, *J. Mol. Struct. (Theochim.)*, 1986, Vol. 136, No. 3–4, p. 371.
146. D.G. Musaev and O.P. Charkin, *Zh. Neorgan. Khimii* (Russ.), 1991, Vol. 36 (to be published).
147. T.S. Zjubina and O.P. Charkin, *Zh. Neorgan. Khimii* (Russ.), 1990, Vol. 35 (to be published).
148. P.G. Jasien and W.J. Stevens, *J. Chem. Phys.*, 1985, Vol. 83, No. 16, p. 2984.
149. W.K. Li, J. Baker and L. Radom, *Austral. J. Chem.*, 1986, Vol. 39, No. 6, p. 913.
150. W. Koch and G. Frenking, *J. Phys. Chem.*, 1987, Vol. 91, No. 6, p. 49.
151. D.G. Musaev, V.V. Jackobson and O.P. Charkin *Koordin. Khimiya* (Russ.), 1989, Vol. 15, No. 8, p. 1011.
152. D. Poppinger and L. Radom, *J. Amer. Chem. Soc.*, 1978, Vol. 100, No. 11, p. 3674.
153. A.D. Molean, G.H. Loew and D.S. Berkowitz, *J. Mol. Spectrosc.*, 1977, Vol. 64, No. 2, p. 184.
154. H. Leung, R. Suffolk and J. Watts, *Chem. Phys.*, 1986, Vol. 109, No. 2,3, p. 277.
155. M.K. Chripun, A.Yu. Efimov, L.S. Lylich et al., *Zh. Strukt. Khimii* (Russ.), 1986, Vol. 27, No. 6, p. 92.
156. D.G. Hopper, *J. Chem. Phys.*, 1984, Vol. 80, No. 9, p. 4290.
157. A.A. Korkin, A.M. Mebel and E.V. Botysov, *Izvestiya Akad. Nauk of Russia, Ser. Khim.* (Russ.), 1988, Vol. 4, p. 900.
158. J.K. Burdett and C.J. Marsden, *New. J. Chem.*, 1988, Vol. 12, No. 10, p. 797.
159. R. Tian, V. Balaji and J. Michl, *J. Amer. Chem. Soc.*, 1988, Vol. 110, No. 21, p. 7225.
160. T.P. Hamilton and H.F. Schaefer, *Chem. Phys. Lett.*, 1990, Vol. 166, No. 3, p. 303.

161. G.M. Chaban, N.M. Klimenko and O.P. Charkin, *Izvestiya Akad. Nauk Russia, Ser. Khim.* (Russ.), 1990, No. 4, p. 794.

162. G.M. Chaban, N.M. Klimenko and O.P. Charkin, *Izvestiya Akad. Nauka of Russia, Ser. Khim.* (Russ.), 1990, No. 7, p. 1590.

163. G.M. Chaban, N.M. Klimenko and O.P. Charkin, *Zh. Neorgan. Khimii* (Russ.), 1990, Vol. 35 (to be published).

164. G.M. Chaban, N.M. Klimenko and O.P. Charkin, *Zh. Neorgan. Khimii* (Russ.), 1990, Vol. 35 (to be published).

165. W.G. Laidlaw and M. Trsic, *Can. J. Chem.*, 1985, Vol. 63, No. 7, p. 2044.

166. W.L. Feng, O. Novaro and J. Garcia-Prieto, *Chem. Phys. Lett.*, 1984, Vol. 111, No. 3, p. 297.

167. F. Moscardo, R. Andarias and E. San-Fabian, *Int. J. Quant. Chem.*, 1988, Vol. 34, No. 4, p. 375.

168. H. Basch, *Chem. Phys. Lett.*, 1989, Vol. 157, No. 1,2, p. 129.

169. T.H. Dunning et al., *J. Phys. Chem.*, 1981, Vol. 85, p. 1350.

170. T. Fueno and R.J. Buenker, *Theor. Chim. Acta*, 1988, Vol. 73, No. 2,3, p. 123.

171. S.P. Konovalov and V.G. Solomonik, *Zh. Strukt. Khimii* (Russ.), 1984, Vol. 25, No. 6, p. 11.

172. F. Ramondo, *Chem. Phys. Lett.*, 1989, Vol. 156, No. 4, p. 346.

173. S. Nakamura, __ Takahaschi, R. Okazaki and K. Morokuma, *J. Amer. Chem. Soc.*, 1987, Vol. 109, No. 14, p. 4142.

174. L.L. Lohr and R.C. Boechm, *J. Phys. Chem.*, 1987, Vol. 91, No. 12, p. 3203.

175. C.E. Ehrhart and R. Ahlrichs, *Chem. Phys.*, 1986, Vol. 108, No. 3, p. 417.

176. M. Nonella, J.R. Huber and T.K. Ha, *J. Phys. Chem.* 1987, Vol. 91, No. 20, p. 5207.

177. N.M. Klimenko, In "Chemical Bonding and Structure of Molecules". Nauka Publ., Moscow, 1984, p. 36.

178. V.V. Jackobson, D.G. Musaev and O.P. Charkin, *Zh. Neorgan. Khimii* (Russ.), 1991, Vol. 36 (to be published).

179. G.H. Yates and R.M. Pitzer, *J. Chem. Phys.*, 1977, Vol. 66, p. 3592.

180. J. Apeloig, P.v.R. Schleyer, J.S. Binkley, J.A. Pople et al., *Tetrahedr. Lett.*, 1976, p. 3923.

181. M.T. Nguyen, *Chem. Phys. Lett.*, 1983, Vol. 97, p. 503.

182. P.S. Burgers, J.L. Holmes and J.K. Terlow, *J. Amer. Chem. Soc.*, 1984, Vol. 106, No. 10, p. 2762.

183. L.L. Lohr, *J. Phys. Chem.*, 1984, Vol. 88, pp. 2603–2992.

184. K. Ito and S. Nagase, *Chem. Phys. Lett.*, 1986, Vol. 126, No. 6, p. 531.

185. T. Kudo and S. Nagase, *J. Amer. Chem. Soc.*, 1985, Vol. 107, No. 9, p. 2589.

186. C.J. Marsden, *J. Chem. Soc. Dalton Trans.*, 1984, No. 7, p. 1279.

187. V.G. Solomonik, *Koordin. Khimiyaa* (Russ.), 1990, Vol. 16 (to be published).

188. V.V. Jackobson, D.G. Musaev and O.P. Charkin, *Zh. Neorgan. Khimii* (Russ.), 1991, Vol. 36 (to be published).
189. T.S. Zjubina and O.P. Charkin, *Zh. Neorgan. Khimii* (Russ.), 1990, Vol. 35 (to be published).
190. T.P. Pogvebnaja and V.G. Solomonik, In "Structure and Molecular Properties", Ivanovo Institute of Chem. Techn., Ivanovo (Russ.), 1988, p. 32.

Table 1

Geometry and energy characteristics[a] of special points of PES of the Li^+AB molecules and ions with 10 valent electrons [87–89].

Molecule Structure[b]		$\varphi(LixA)$ degree	R Å	r_{AB} Å	E_{rel}^{c}, kcal/mol		Refs.
					SCF	MP 3	
LiCN,	I	0	1.91	1.16	5.8	2.1	87
	II	56	2.18	(1.16)	8.6	5.5	"
CNLi,	V	180	1.77	(1.16)	0	0	"
experim.		180	1.77	1.17			39
LiBO,	I	0	2.26	1.22	0	0	87
	II	60	2.37	1.22	11.9	7.8	"
BOLi,	V	180	1.56	1.27	-2.7	0.7	"
LiSiN,	I	0	2.4	1.55	62.7		89
	II	45	52.86	1.55	62.7		"
SiNLi,	V	180	1.75	1.56	0		"
LiAlO,	I	0	2.61	(1.56)	55.7		"
	II	69	2.72	1.58	63.7		"
ALOLi,	V	180	1.64	1.61	0		"
Li^+CO,	I	0	2.27	1.13	0	0	87
	II	77	2.79	(1.13)	12.0	13.2	"
$COLI^+$	V	180	1.91	(1.13)	-1.6	3.5	"
LiCP,	I	0	1.96	1.54	0		88
CPLi,	V	180	2.30	1.59	39.6		"
LiBS,	I	0	2.19	1.62	0		"
	II	60	2.37	(1.65)	11.5		"
	III	80	1.87	1.67	10.6		"
BSLi,	V	180	2.19	(1.72)	25.4		"
LiSiP,	I	0	2.48	1.94	16.8		89
	II	38	3.15	1.94	18.8		"
SiPLi,	V	109	2.49	1.99	0		"
	V	180	2.29	1.97	4.4		"
LiAlS,	I	0	2.64	(2.05)	23.3	22.0	89
	II	67	2.79	2.07	32.2	29.5	"
AlSLi,	IV	120	2.44	2.16	0	0	"
	V	180	2.16	2.15	1.8	2.8	"
Li^+CS,	I	0	2.15	1.48	0		88
$CSLi^+$,	V	180	2.73	1.53	29.3		"
$LiBF^+$,	V	0	2.39	1.28	0	0	87
	III	87	2.72	1.34	26.4	23.3	"
$BFLi^+$,	V	180	1.77	1.39	18.6	16.8	"

Table 1 (continued)

Molecule Structure[b]		$\varphi(\text{LixA})$ degree	R Å	r_{AB} Å	E_{rel}^{c}, kcal/mol		Refs.
					SCF	MP 3	
LiSiO⁺,	I	0	2.86	1.47	57.8		89
	III	63	3.34	1.48	60.0		"
SiOLi⁺,	V	180	1.73	1.51	0		"
LiAlF⁺,	I	0	2.80	1.58	22.5	25.0	89
	III	82	2.90	1.63	33.5	33.1	"
AlFLi⁺,	V	180	1.68	1.71	0	0	"
Li⁺NN,	I	0	2.10	1.09	0	0	87
	III	90	2.38	1.09	9.6	10.9	"
LiBCl⁺,	I	0	2.35	1.63	0	0	88
	III	97	2.78	1.68	29.5	27.6	"
BClLi⁺,	IV	152	2.45	1.82	24.1	23.7	"
	V	180	2.52	1.73	28.2	27.9	"
LiSiS⁺,	I	0	2.79	1.87	19.6		89
	II	56	3.39	1.89	26.0		"
SiSLi⁺,	V	180	2.40	1.94	0		"
LiAlCl⁺,	I	0	2.82	2.06	13.6		"
	II	72	3.19	2.16	27.3		"
AlClLi⁺,	V	180	2.26	2.39	0		88
Li⁺NP,	I	0	1.90	1.45	0		"
NPLi⁺,	V	180	2.89	(1.46)	47.5		89
Li⁺PP,	I	0	2.62	1.84	0		"
	III	90	2.72	1.86	4.0	1.4	"

[a]Geometry was optimized within the SCF/3-21G* approximation; the relative energies were precized within the SCF and MP3 approximations using the DZHD + P basis set [87–89].
[b]By Roman figures the structures of Figure 2 are denoted.
[c]RElative energies above the ground isomer.

Table 2

Geometry and energy characteristics of special points of PES of the Cu^+ complexes with the AB molecules and AB anions containing 10 valent electrons

Molecule Structure[b]	φ(CuxA) degree	R Å	r_{AB} Å	E_{rel}[c], kcal/mol	$D(Cu^+)$ kcal/mol	Refs.
CuCN, I	0	1.91	1.15	0	162	91,92
II	68	2.12	(1.16)	17.8		"
CNCu, V	180	1.85	1.16	0.7		"
CuBO, I	0	2.04	(1.25)	0	175	92
II	69	2.30	(1.25)	22.8		"
BOCu, V	180	1.83	(1.25)	10.7		"
CuBeF, I	0	2.24	1.42	0	178	"
II	75	2.09	1.46	24.5		"
BeFCu, V	180	1.81	1.52	42.9		"
CuCP, I	0	1.94	(1.60)	0		93
CPCu, V	180	2.27	(1.60)	43.8		"
CuAlO, I	0	2.43	(1.59)	31.5		"
II	62	2.67	(1.59)	46.9		"
AlOCu, V	180	1.82	(1.59)	0		"
Cu^+N_2, I	0	2.01	1.10	0	23.0	94
III	90	2.26	1.10	31.6		"
Cu^+CO, I	0	2.06	1.12	0	17.1	91,92
II	83	2.32	1.13	22.8		"
$CoCu^+$, V	180	2.00	1.15	0.7		"
Cu^+BF, I	0	2.18	(1.31)	0		"
III	95	2.53	(1.31)	32.3	30.6	"
$BFCu^+$, V	180	2.09	(1.31)	15.7		91,92
Cu^+CS, I	0	2.08	(1.52)	0	37.7	95
$CSCu^+$, V	180	2.45	(1.52)	35.6		"
Cu^+P_2, I	0	2.48	1.86	1.3		94
III	90	2.49	1.89	0		"

[a] Geometry was optimized with the use of the $(DZRVV)_{Cu}$ + $(DZRS)_{A,B}$ basis set (values r_{AB} in brackets were fixed), while the energies with the $(DZW)_{Cu}$ + $(DZHD)_{A,B}$ basis [19,90–92]. The CuCP, CuOAl and CuP_2^+ systems were calculated using the pseudopotential method [93,94]. All other molecules and ions were calculated within the all-electron SCF approximation.

[b] Structures I—V are shown in Figure 2.

[c] Relative energies above the ground isomer.

[d] Energy of decay $CuAB \to Cu^+ + AB^-$ and $CuAB^+ \to Cu^+ + AB$ producing the Cu^+ cation, were calculated with the superposition basis set effects taken into account. The $CuCS^+$ in exception, and evaluation of its $D(Cu^+)$ can be overestimated by 5–10 kcal/mol.

Table 3

Geometry and energy characteristics[a] of special points of PES of molecules and ions of M^{q+} cations with diatomic AB ligands containing 10 valent electrons

Molecule Structure[b]	$\varphi(MxA)$ degree	R Å	r_{AB} Å	E_{rel}[c], kcal/mol		$D(M^{q+})$, kcal/mol		Refs.
				SCF	MP 3	SCF	MP 3	
$Be^{2+}CO$, I	0	1.80	1.09	0	0	92	99	122,123
III	86	1.83	1.12	43	43			"
$COBe^{2+}$, V	180	1.51	1.17	0	11			"
$Be^{2+}CS$, I	0	1.72	1.45	0		161		126
$CSBe^{2+}$, V	180	1.93	1.59	92				"
$Be^{2+}NP$, I	0	1.53	1.46	0		200		"
$NPBe^{2+}$, V	180	2.13	1.47	142				"
$BeCN^+$,I	0	1.61	1.14	17.5	9			"
II	49	1.90	1.15	25.1	15			"
$CNBe^+$, V	180	1.45	1.18	0	0	400		"
NaCN, I	0	2.24	1.17	3.6				31
CNNa, IV	104	2.27	1.15	-0.8				"
experim.	96	2.23	1.17					34
V	180	2.09	1.17	0				31
KCN, I	0	2.69	1.16	6.1				31,32
CNK, IV	102	2.67	1.15					"
experim.	91	2.52	1.16					35
V	180	2.51	1.17					31,38
Al^+CO, I	0	3.26	1.10	0.2				128,129
$COAl^+$, V	180	2.89	1.11	0		4.1		"
CNRb, IV	109	2.82	1.16	0				36
V	180	2.54	1.16	1.2				"
BeCN, I	0	1.69	1.13	9	6			120
III	(90)	1.70	1.14	21.4	16.4			"
CNBe, V	180	1.53	1.16	0	0	94.5	94.3	120
MGCN, I	0	2.09	1.14	6.7	3.0			"
III	(90)	2.11	1.14	8.4	5.3			"
CNMG, V	180	1.93	1.15	0	0	75.5	77.3	"
CaCN,I	0	2.42	1.14	7.8	4.3			"
III	(90)	2.41	1.14	5.5	3.0			"
CNCa, V	180	2.23	1.15	0	0	95.0	94.0	"
BaCN, I	0	2.74	1.14	7.4				"
III	(90)	2.75	1.14	3.9				"
CNBa, V	180	2.55	1.15	0	0	100.1	99.9	"
Na^+CO, I	0	2.65	1.09	0.5				125
$CONa^+$, V	180	2.31	1.11	0		8.7		"
$Mg^{2+}CO$,I	0	2.24	1.11	0	0	43.5	49.2	123
$COMg^{2+}$,V	180	1.90	1.17	2.1	12.2			"
$Al^{3+}CO$, I	0	2.06	1.08	0				127
II	56.6	2.24	1.13	56.6				"
$COAl^{3+}$, V	180	1.74	1.20	-4.5				"

[a]In [123,126] geometry was optimized with the use of the 3-21G* basis set, and energies with the DZHD + P. The optimization for the Na and K cyanides were carried out with the DZ + 2P basis containing double-zta polarizing 3d functions on the C and N atoms [31–33], while for the alkali-earth cyanides with more flexible than TZ + 2P basis set. The energies were precized with correlation energy taken into account within the CISD approximation [120]. For the Al^+ and Al^{3+} carbonyles the DZ + P basis set was used [127–129].

[b]Structures I–V are drawn on Figure 4.

[c]Energy of decay $M^{q+} \cdot AB \rightarrow M^{q+} + AB$ or $M^{q+} \cdot AB^- \rightarrow M^{q+} + AB^-$.

Table 4

Geometry and energy characteristics[a] of special points of PES of the rearrangements $XM \cdot CO \to XM \cdot OC$ and $XM \cdot NP \to XM \cdot PN$ where X = F, H and Li.

A. Compounds of Li^+ [123,126]

Molecule Structure[b]	$\varphi(LixA)$ degree	$R(Li-X)$ Å	R Å	$r(A-B)$ Å	α de-gree	E_{rel}, kcal/mol SCF	E_{rel}, kcal/mol MP 3	$D(Li-AB)$[c], kcal/mol SCF	$D(Li-AB)$[c], kcal/mol MP 3
FLiCO I	0	1.51	2.26	1.10		0	0	8.5	10.7
II	77.3	1.51	2.79	1.13		8.7	9.2		
FLiOC III	180	1.52	2.00	1.12		0.5	1.9		
HLiCO I	0	1.64	2.30	1.10		0	0	8.3	9.9
II	76.8	1.64	2.90	1.13		8.3	7		
HLiOC III	180	1.64	2.00	1.12		0.5	1.8		
LiLiCO I	0	2.84	2.36	1.10		0	0	3.5	5.6
II	76.5	2.80	3.82	1.13		4.0	3.4		
LiLiOC III	180	2.83	2.03	1.12		-0.3	0.7		
$H_2Li \cdot CO$ Ia	0	1.79	2.59	1.10	166	0	0	1	
$H_2Li \cdot OC$ IIIa	180	1.78	2.90	1.12	173	0	0		
HLiNP I	0	1.66	202	1.14		0		26	
HLiPN III	180	1.63	3.04	1.46		26			
LiLiNP I	0	2.88	2.01	1.45		0		21	
LiLiPN III	180	2.75	3.31	1.46		decay[d]			

A. Compounds of Li$^+$ [123,126]

Molecule[b] Structure	φ(CuxA) degree	R(Cu-X) Å	R Å	r(A-B) Å	α degree	E$_{rel}$, kcal/mol SCF	D(Cu-AB)[c], kcal/mol SCF
FCuCO I	0	(1.75)	1.87	(1.12)		0	22.3
II	(83)	(1.75)	2.16	(1.13)		29.6	
FCuOC III	180	(1.75)	1.92	(1.15)		7.4	
HCuCO I	0	(1.47)	1.98	1.12		0	18.3
II	(83)	(1.47)	2.42	1.13		24.4	
HCuOC III	180	(1.47)	1.99	1.16		5.1	
FCuCN I	0	(1.75)	1.91	(1.15)		1.4	40.1
II	(68)	(1.75)	2.03	(1.16)		24.2	
FCuNC III	180	(1.75)	1.97	(1.16)		0	
HCuCN I	0	(1.47)	1.98	(1.15)		1.3	
II	(68)	(1.47)	2.08	(1.16)		22.3	
HCuNC III	180	(1.47)	1.91	(1.16)		0	37.4
H$_2$CuCN^{2-} Ia	0	(1.47)	2.19	(1.16)	141	0	
H$_2$CuCN^{2-} IIIa	180	(1.47)	2.14	(1.16)	146	0.3	
H$_2$·CO Ia	0	(1.47)	1.90	(1.12)	130	0	
H$_2$Cu·OC IIIa	180	(1.47)	2.15	(1.15)	166	8.4	

Geometry of the Li$^+$ compounds was optimized on the SCF/3-21G level, the energies were precized with the DZHD + P basis set. Geometry optimization of the Cu$^+$ compounds was carried out with the (DZRVV)$_{Cu}$ + (DZRS)$_{C,O,N}$, while the energetic characteristics—with the (DZW)$_{Cu}$ + (DZHD)$_{C,O,N}$ basis sets [130].

b Structures I—III and Ia—IIIa are shown on Figure 4. The φangle and R distance are taken in the same manner as in the triatomic systems MAB (see Figure 2). α is angle XMX in the X$_2$M · AB systems.

c Energy decay with separation of AB ligand.

d Structure LiLiPN (III) is not stable to decay Li$_2$ + PN.

Table 5

Geometry and energy characteristics[a] of the hydrated systems (Li⁺ · H₂O)AB

Molecule	Structure[b]	φ(LixA) degree	R(LiO) Å	R Å	rAB[c] Å	E_{rel} kcal/mol	D(Li-H₂O)[d] kcal/mol	D(Li-AB)[e] kcal/mol
H₂OLi⁺CO	Ia		1.794	2.255	(1.128)	12.2		
	Ia	72	1.796	2.587	(1.128)	0	33.5	13.4
H₂OLi⁺OC	IIIa		1.797	1.914	(1.128)	0		
H₂OLi⁺NP	Ia		1.811	1.936	(1.451)	45.2	29.9	38.7
H₂OLi⁺PN	IIIa		1.785	2.889	(1.459)	2.1		
H₂OLiBO	Ia		1.833	2.193	(1.226)	12.0		
	IIa	61	1.823	2.33	(1.208)	0.0	27.4	140.8
H₂OLiOB	IIIa		1.842	1.654	(1.286)	5.4		
H₂OLiCN	Ia		1.835	1.972	(1.157)	0		
H₂OLiNC	IIIa		1.841	1.785	(1.157)	0	22.3	144.2
H₂OLiCP	Ia		1.841	1.982	(1.544)	36.3	21.5	142.7
H₂OLiPC	IIIa		1.818	2.326	(1.587)	56.5		
H₂OLiSiN	Ia		1.818	2.472	(1.546)	0		
H₂OLiNSi	IIIa		1.852	1.774	(1.546)	12.9	19.2	157.6
H₂OLiSiP	Ia	72	1.820	2.506	1.940	0		
	IIa		1.827	2.524	1.985		22.3	122.1
H₂OLiPSi	IIIa		1.827	2.327	1.969	2.0		

[a] Geometry was optimized with the 3-21G* basis set, energies were calculated within the SCF/DZHD + P approximation. Geometry the H₂O molecule was fixed R(OH) = 0.965 Å; < HOH = 110°).

[b] Strcutres Ia-IIIa are drawn on Figure 4.

[c] Date in brackets were fixed and taken from the SCF/3-21G* optimization of LiAB triatomics.

[d] Energy of reaction H₂OLi⁺AB → H₂O + Li⁺AB;

[e] Energy of reaction H₂OLi⁺AB → AB + Li⁺ · H₂O.

Table 6

Geometry and energy characteristics[a] of altenrative structures of the LiAB molecules with 12 valent electrons [133].

Molecule Structure[b] term		φ(LixA) degree	R Å	r_{AB} Å	E_{rel}, kcal/mol SCF	MP 3
LiNO I	S	0	1.69	1.18	15	12
	T	0	1.70	1.19	-25	-12
II	S	47	2.27	1.21	21	20
	T	46	2.14	1.22	-21	-8
III	S	95	1.65	1.23	0	0
	T	100	1.72	1.26	-32	-15
IV	S	151	2.17	1.26	24	30
	T	144	2.10	1.28	-30	-10
V	S	180	1.59	1.24	21	26
	T	180	1.58	1.27	-32	-11
LiNs I	S	0	1.73	1.54	10	9
	T	0	1.76	1.56	-28	-13
II	S	28	1.78	1.55	13	13
	T	35	1.78	1.57	-26	-10
III	S	76	1.89	1.57	0	0
	T	72	1.96	1.62	-29	-11
V	S	180	2.09	1.57	37	42
	T	180	2.08	1.66	-13	13
LiPO I	S	0	2.16	1.48	36	35
	T	0	2.15	1.48	6	16
II[c]	S	62	2.78	1.49	20	18
	T	47	2.79	1.50	4	13
III	S	113	1.87	1.54	0	0
	T					
V	S	180	1.60	1.56	13	18
	T	180	1.60	1.57	-29	-13
LiPS I	S	0	2.23	1.92	33	34
	T	0	2.24	1.93	5	18
III	S	94	2.11	1.98	0	0
	T	104	2.26	2.03	-17	-3
V	S	180	2.13	2.00	29	35
	T	180	2.12	2.03	-9	9

[a]Geometry was optimized on the SCF/3-21G level, the energies were calculated within the SCF and MP3 approximations using the DZHD + P basis set.

[b]Structures I-V see on Figure 2.

[c]II corresponds to very shallow section of PES between the linear I and the cyclic III structures of LiPO.

Table 7

Geometry and energy characteristics[a] of alternative structures of the LiAB molecules and LiAB$^+$ ions with 14 valent electrons [136].

Molecule Structure[b]		φ(LixA) degree	R Å	r_{AB} Å	E_{rel}, kcal/mol
LiOF	I	0	1.54	1.36	16
	III	89	1.51	1.48	0
OFLi	V	180	1.53	1.60	24
LiOCl	I	0	1.62	1.56	3
	III	66	1.84	1.71	0
OClLi	V	180	2.08	2.10	44
LiFF$^+$	I	0	1.99	1.35	2.6
	II	53	2.32	1.35	0
III(C_{2v})		90	2.05	1.35	1.2
LiClF$^+$	I	0	3.77	1.58	1.8
	VI	162	1.72	1.63	0
LiFCi$^+$	V	180			0.2
LiSF	I	0	2.02	1.59	27
	III	111	1.77	1.73	0
	V	180	1.56	1.77	20
LiSCl	I	0	2.09	2.03	24
	III	92	2.02	2.19	0
	V	180	2.11	2.35	33
LiClCl$^+$	I	0	2.55	2.01	9
	II	68	2.46	2.02	0
	III	90	2.37	2.02	0

[a]Geometry was optimized on the SCF/3-21G* level, energies were calculated within the SCF/ DZHD+P approximation. Bond energies D(Li$^+$—AB) in ions calculated are (in kcal/mol) 6 for Li$^+$F$_2$, 18 for Li$^+$FCl and 10 for Li$^+$Cl$_2$.
[b]Structures I—IV see on Figure 2.

Table 8
Geometry parameters and relative energies[a] of special points of PES of the MYAY' molecules with
16 valent electrons.

A. Compounds of Li[+] [141,142]

Molecule, structure[b]		R$_e$, Å			Angles, degree		E$_{rel}$ kcal/mol
		LiY,LiY'	AY	AY'	YAY'	LiYA (LiY'A)	
LiOB	I	1.61	1.28	1.21	180	180	0
	III	1.90	1.25	1.25	149	68	139
LiOBS	I	1.64	1.26	1.65	180	180	0
	III						
LiSBO	VI	2.14	1.20	1.72	180	180	24
LiSBS	III	2.24	1.66	1.66	160	60	0
	V						
	VI	2.16	1.70	1.62	180	180	15.1
LiOAlO	I	1.61	1.62	1.57	180	180	0
	III	2.16	1.61	1.61	119	74.6	23.2
LiOAlS	I	1.62	1.61	2.00	180	180	0
LiSAlO	IV	2.73	1.56	2.08	183	89	35.7
	VI	2.16	1.56	2.07	180	180	37
	V	3.39	1.56	2.08	175	104	34
LiSAlS	III	2.56	2.04	2.04	135	65	0
	V	3.48	2.09	2.00	184	107	0.7
	VI	2.17	2.07	2.00	180	180	5.3
	IV	3.08	2.09	2.00	190	90	0.9
Li[+]OCO	I	1.80	1.16	1.12	180	180	0
	III	2.48	1.14	1.14	170	64	25.1
Li[+]OS	I	1.79	1.16	1.52	180	180	0
Li[+]SCO	V	2.52	1.12	1.60	180	106	10.6
	VI	2.57	1.11	1.58	180	180	17.0
	IV	2.54	1.13	1.56	180	71	21.4
Li[+]SCS	III	2.44	1.54	1.54	176	59	7.0
	V	2.50	1.58	1.51	(180)	104	0
	VI	2.49	1.57	1.50	180	180	6.3
Li[+]OSiO	I	1.74	1.50	1.42	180	180	0
	III	2.43	1.49	1.49	136	75.6	47.7
Li[+]OSiS	I	1.73	1.51	1.86	180	180	0
Li[+]SSiO	V	2.47	1.47	1.92	180	126	33.0
	VI	2.46	1.47	1.90	180	180	34.9
	IV	2.90	1.48	1.88	180		50
Li[+]SSiS	III	2.87	1.90	1.90	154	66	15.5
	V	2.45	1.93	1.87	180	125	0
	VI	2.44	1.92	1.87	188	180	2.0

Table 8 (continued)
B. Compounds of Cu[+] [146]

Molecule, structure[b]		R$_e$, Å			Angles, degree		E$_{rel}$ kcal/mol
		LiY,LiY′	AY	AY′	YAY′	LiYA (LiY′A)	
CuOBO	I	1.80	1.30	1.24	180	180	0
	III	2.02	1.27	1.27	157	66	28
CuOBS	I	1.83	1.28	1.68	180	180	0
	VI	2.20	1.23	1.73	180	0	28
CuSBS	I	2.22	1.75	1.66	180	180	16.4
	III	2.25	1.71	1.71	164	58	0
CuOAlO	I	1.78	1.64	1.66	180	180	0
	III	2.39	1.65	1.65	122	79	26
CuSAlS	I	2.22	2.11	2.08	180	180	7.3
	III	2.56	2.12	2.12	133	65	0

[a]Geometry of the Li compounds was optimized within the SCF/3-21G[*] and energies SCF/DZHD + P approximations. Geometry of the Cu compounds was optimized with the (DZRVV)$_{Cu}$ + (DZRS)$_{A,Y,Y′}$ and energies—with the (DZW)$_{Cu+}$ + (DZHD)$_{A,Y,Y′}$ basis sets, correspondingly.
[b]Structures I—VII are drawn on Figure 7. In the "substituted" MYAY′ systems Y and Y′ are more electronegative and more electropositive chalkogen, correspondingly, A—central atom of the YAY′ ligand.

Table 9

Geometry and energy characteristics[a] of special points of PES of the MNCS and MCNS molecules (M^+ = Li^+ and Cu^+)

A. Compounds of Li^+ [151]

Molecule, structure[b]	R_e, Å			Angles, degree		E_{rel} kcal/mol
	MY,MY'	AY	AY'	YAY'	MYA (MY'A)	
LiNCS IV	1.76	1.17	1.68	180	180	0
V	2.10	1.16	1.71	171	71	9.1
VI	2.26	1.15	1.76	170	66	5.6
NCSLiVIII	2.08	1.15	1.71	180	0	22
LiCNS IX	1.96	1.15	1.77	180	180	14.3
X	2.59	1.15	1.79	172	111	18.7
XI	2.04	1.17	1.74	176	58	0
CNSLiXIII	2.07	1.16	1.70	180	0	19.5

B. Compounds of Cu^+ [146]

CuNCS IV	1.87	(1.17)	1.74	180	180	0
V	2.18	(1.17)	1.79	180	80	24
VI	2.50	(1.17)	1.82	180	73	16
NCSCuVIII	2.38	(1.17)	1.74	180	0	41
CuCNS IX	1.97	(1.16)	1.78	180	180	0
X	2.40	(1.16)	1.77	180	95	17
XI	2.31	(1.16)	1.74	180	68	0
CNSCuXIII	2.24	(1.16)	1.71	180	0	28

[a]Geometry of the Li compounds was optimized with the 3-21G*, and of the Cu compounds with the $(DZRVV)_{Cu,N,S}$ basis sets. Energies of special points of PES were precized with the DZHD + P and $(DZW)_{Cu}$ + $(DZHD)_{C,N,S}$ basis sets correspondingly.
[b]Structures IV—XIII are drawn on Figure 9.

Table 10

Geometry and energy characteristics[a] of the triatomic A_2Y and ABY molecules with 16 valent electrons [161]

A. Molecules N_2O, N_2S, P_2O, P_2S, ONP and SNP

Molecule Structure[b] symmetry	R_{YA} Å	R_{YB} Å	R_{AB} Å	Angles[c], degree			E_{rel}^{c}, kcal/mol	
				YAB	ABY	AxY	SCF	MP 3
N_2O I $C_{\infty v}$	1.18		1.09	180			0	0
III C_{2v}							69	56
N_2S I $C_{\infty v}$	1.65		1.09	180			0	
III C_{2v}	2.72		1.08	78		90	26	
P_2O I $C_{\infty v}$	1.44		1.84	180			0	0
II C_s	1.46		2.04	(120)		144	22	
II' C_s	1.48		2.07	(90)		125	23	33
III C_{2v}	1.71		1.93	56		90	8	4
P_2S I $C_{\infty v}$	1.89		1385	180			6	5
II C_s	1.94		1.97	(120)		139	26	31
II' C_s	1.98		1.97	(90)		116	21	
III C_{2v}	2.16		1.95	63		90	0	0
PSPVI $D_{\infty h}$	1.87						68	
OPNI $C_{\infty v}$	1.44		1.45	180		0	5	7
II C_s	1.47		1.61	87		64	32	53
III C_s	1.59	1.61	1.53	62	61	89	32	32
PNOV C_s	1.88	1.21	1.70	180	107	136	54	73
V $C_{\infty v}$	1.93	1.16	1.49	(120)	180	180	0	0
SPNI $C_{\infty v}$	1.94	1.45		(105)		0	27	32
II C_s	1.96	1.54		(90)		44	47	
II' C_s	2.10	1.58		59		55	48	70
II' C_s		1.55			76	68	47	
III C_s		1.53	1.85		(90)	100	24	29
IV C_s		1.52	1.76		(120)	113	32	
IV' C_s		1.51	1.72		180	137	41	50
PNSV $C_{\infty v}$		1.48	1.58			180	0	0
PSNVI $C_{\infty v}$	1.84		1.43				91	86

[a]Geometry was optimized with the 3-21G* and energies were precized with the DZHD + P basis sets correspondingly.
[b]Structures I—VI are drawn on Figure 12. Y is chalkogen (O and S), A—phosphorus, B—nitrogen (in nonsymmetrical molecuels ABY). The AxY angle has its vertex in the middle of the A—B bond.
[c]Values of angles in the brackets are fixed, all other geometric parameters were optimized.

Table 10 (continued)
B. Anions N_3^-, N_2P^-, NP_2^- and P_3^- [162]

Anion Structure[a] Symmetry			R_{12}	R_{23}	φ	θ	E_{rel} kcal/mol	
							SCF	MP 3
N_3^-	I	$D_{\infty h}$	1.15	1.15	180	0	0	0
	II	C_{2v}	1.24	1.24	(120)	41	91	
	II'	C_s	1.20	1.35	(105)	54	102	
	II"	C_s	1.22	1.28	(90)	65	80	
	III	D_{3h}	1.37	1.37	(60)		119	
	III'	C_{2v}	1.55	1.55	45	90	77	73
P_3^-	I	$D_{\infty h}$	1.91	1.91	180	0	0	0
	II	C_{2v}	2.01	2.01	(120)	41	39	
	II'	C_{2v}	2.00	2.00	(90)	63	20	
	II"	C_{2v}	2.03	2.03	73	78	9	10
	III"	C_{2v}	2.08	2.08	(65)		14	
	III	D_{3h}	2.13	2.13	(60)		24	
	III'	C_{2v}	2.15	2.15	(58)		20	23
	III'	C_{2v}	2.27	2.27	51	90	10	16
P_2N^-	I	$D_{\infty h}$	1.56	3.12	0	180	0	0
	II	C_s	1.61	2.82	(30)	139	53	
	III	C_3	1.60	2.27	51	110	26	38
	IV	C_s	1.52	2.07	(90)	70	50	
	IV'	C_s	1.55	2.00	(120)	44	54	
	V	$C_{\infty v}$	1.47	1.92	(180)	0	8	22
PON_2^-	I	$C_{\infty v}$	1.66	2.77	0		0	0
	IV	C_{2v}	1.56	1.56	72		76	61
	IV'	C_{2v}	1.55	1.55	(90)		80	
	OV'	C_{2v}	1.57	1.57	(120)		77	
	V	$D_{\infty h}$	1.48	1.48	180		33	22

[a]Latin Figures 1 and 3 denote terminal atoms, 2—a central atom. φ is an angle with a vertex at the central atom 2, θ—an angle with a vertex at the middle of the double bond. Computational approximation is similar to that used for N_2O, N_2S etc. (See note after Table 10A).

Table 11

Geometry and energy characteristics[a] of special points of PES of the Li$_2$AB$^+$ ions with 10 valent electrons [178]

Ion, Structure,[b] symmetry	R(Li$_1$–A) Å	R(Li$_2$–B) Å	r(AB) Å	R$_1$ Å	R$_2$ Å	φ_1 degree	φ_2 degree	E_{rel}, kcal/mol
Li$_2$BO$^+$ I C$_{2v}$	2.29		1.19	2.70	2.70	46	46	50[c]
II C$_s$	III	1.71	1.23				180	0
III C$_{\infty v}$	2.26		1.24			151	80	23
IV C$_s$	III		1.26			126	126	21
VIII C$_s$		1.74	1.61	2.26	2.21	0	180	13
VIII C$_{2v}$		1.81	1.62	2.10	2.10	164	73	28
Li$_2$AlO$^+$ I C$_{\infty v}$	2.72	1.65	1.71	2.42	2.77	138	138	0
VIII C$_s$		1.66	1.60	2.26	2.26	40	40	17
VIII C$_{2v}$		1.72	1.65	2.85	2.85	0	180	2
Li$_2$BS$^+$ I C$_{2v}$	2.30		1.66			0	133	0
III C$_{\infty v}$	2.26	2.32	1.71			132	79	22
IV C$_s$	2.27	2.33	1.77		2.36	61	61	22
VIII C$_s$		2.38		2.86	2.82			
VIII C$_{2v}$		2.38		2.68	2.68			
Li$_2$AlS$^+$ I C$_{2v}$	2.84		1.99	3.51	3.51	42	42	66[c]
III C$_{\infty v}$	2.72	2.24	2.06			0	180	10
VII C$_s$		2.27	2.09	3.17	3.09	154	61	22
VIII C$_{2v}$		2.29	2.27	2.97	2.97	136	136	0

[a] Scanning of PES and geometry optimization were carried out with the 3-21G* and relative energies were precized with the DZHD + P basis sets correspondingly.

[b] Structures I—VIII are drawn on Figure 14. R$_1$ and R$_2$ are distances from the Li$_1$ and Li$_2$ atoms to x—the middle of the A—B bond. Angles φ_1 = Li$_1$xA and φ_2 = Li$_2$xA (see notes to Figure 14).

[c] Energies of high-lying structure I of Li$_2$BO$^+$ and Li$_2$AlS$^+$ were calculated with the 3-21G* basis set.

Table 12

Geometry and energy characteristics[a] of alternataive structures of the (LiAB)₂ dimers [186,188].

Molecule, structure[b]	Internuclear distances, Å			Angles, degree						E_{rel} kcal/mol
	r_{AB}	R_{LiA}	R_{LiB}	(LiBLi)	(LiALi)	(ABA)	(ALiA)	(BLiB)	(ALiB)	
(LiBO)₂										
I	1.23	2.36	1.79	80	64					35
II	1.30		1.77							5
IV	1.26	2.31	1.77				86	120	104	0
V	1.26	2.33		121[c]	135[d]					0
VII	1.30		1.64							74[e]
(LiAlO)₂										
I	1.59	2.85		75	56					not stable
II	1.67	2.51	1.77			70				0
IV	1.63		1.68				99	130		60
VII	1.81		1.66			86				20
(LiBS)₂										
I	1.62	2.38	2.38	82	63					11
II	1.78	2.30	2.44							42
IV	1.68		2.36				99	150		0
(LiAlS)₂										
II	2.18	2.78	2.36	74						0
IV	2.08		2.38				89	144	108	24
VII										
(LiCN)₂										
I	1.16	2.09	1.93	77	72					22
II	1.18		1.95							0
IV	1.16	2.07	1.93				99	118		4.5
V	1.16	2.09		118[c]	134[d]					4.5

[a] Geometry of alternative structures for (LiBO)₂ was optimized with 3-21G basis set, for (LiBS)₂, (LiAlO)₂ and (LiAlS)₂—with G$_{AlS}$ (3d polarization functions of Al and Si atoms were included) [188], and for (LiCN)₂—with 4-31G [186] basis sets. Energies were precised with using the DZHD + P [188] and 6-31 + G* [186] basis set.

[b] Structures I—VIII are drawn on Figure 16.

[c] Angle φ(LiBA), and d) angle φ(LiAB) for the structure V.

[e] Using 3-21G basis set.

[f] Using 3-21G + 3d$_{Al}$ basis set.

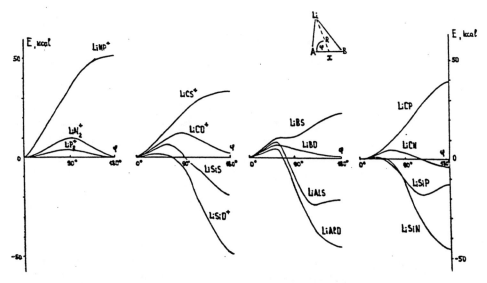

Figure 1. Profiles of PES along the MEP of migration of the Li$^+$ cation around the AB ligands with 10 valent electrons [87–89].

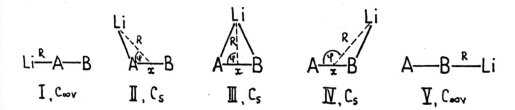

Figure 2. Alternative configurations of the MAB molecules and ions, corresponding to the special points of PES of the MAB ↔ ABM rearrangements.

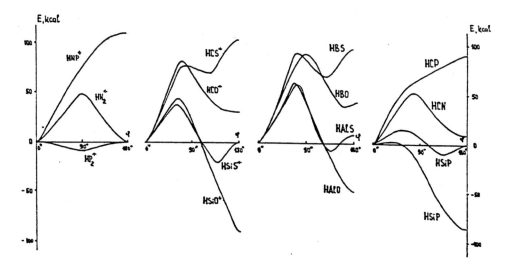

Figure 3. Profiles of PES along the MEP of the 1,2-hydrogen shift rearrangements HAB <-> ABH in systems with 10 valent electrons [126].

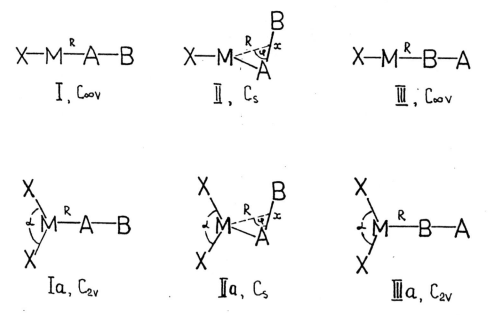

Figure 4. Alternative structures of the XM · AB molecules (—III) and X₂M⁻ · AB ions (Ia—IIIa).

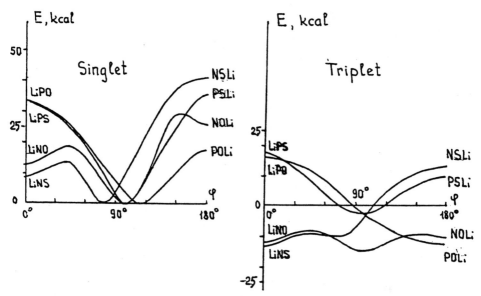

Figure 5. Profiles of PES along MEP of migration of the Li$^+$ cation around the AB$^-$ anions with 12 valent electron [133].

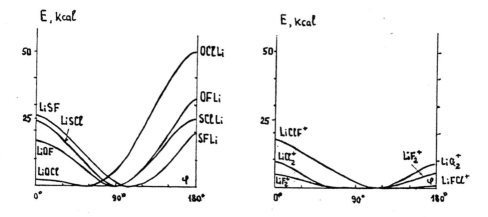

Figure 6. Profiles of PES along MEP of migration of the Li$^+$ cation around the AB$^-$ anion with 14 valent electrons [136].

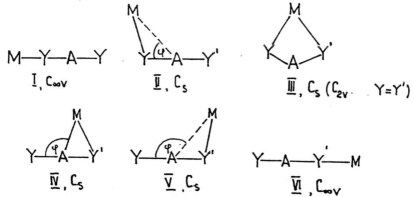

Figure 7. Alternative structures of the MABC molecules.

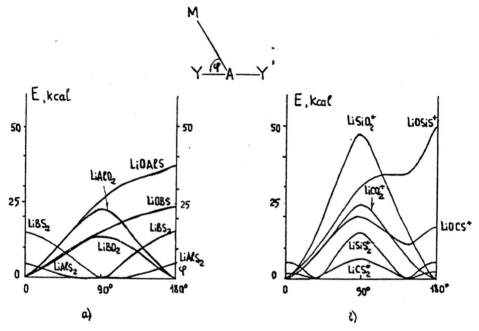

Figure 8. Profiles of PES along the MEP of migration of the Li^+ cation around the YAY^L anions (a) and YAY'(b) molecules with 16 valent electrons [141,142].

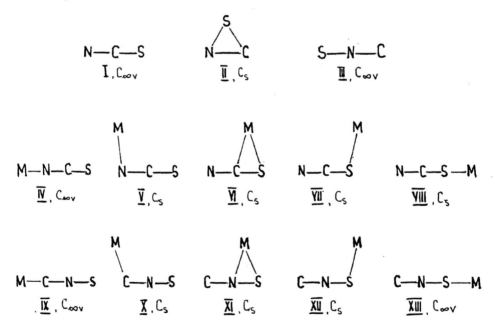

Figure 9. Alternative structures of the NCS⁻ anion (I—III) and its salt molecules MNCs (IV—VIII), and MCNS (IX—XIII).

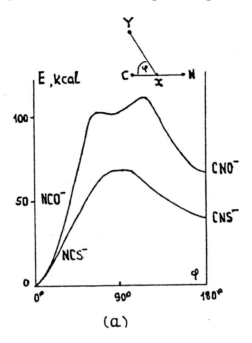

(a)

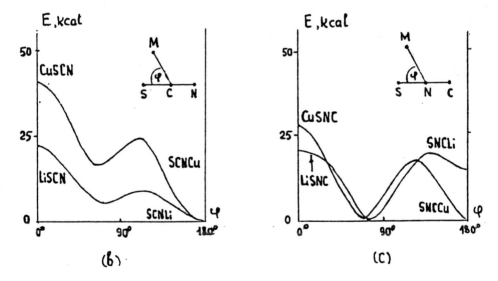

(b)   (c)

Figure 10. Profiles of PES along the MEP of the rearrangements NCO⁻ ↔ CNO⁻ and NCS⁻ ↔ CNS⁻ (a), LiNCS ↔ NCSLi and CuNCS ↔ NCSCu (b), LiCNS ↔ CNSLi and CuCNS ↔ CNSCu (c) [146,151].

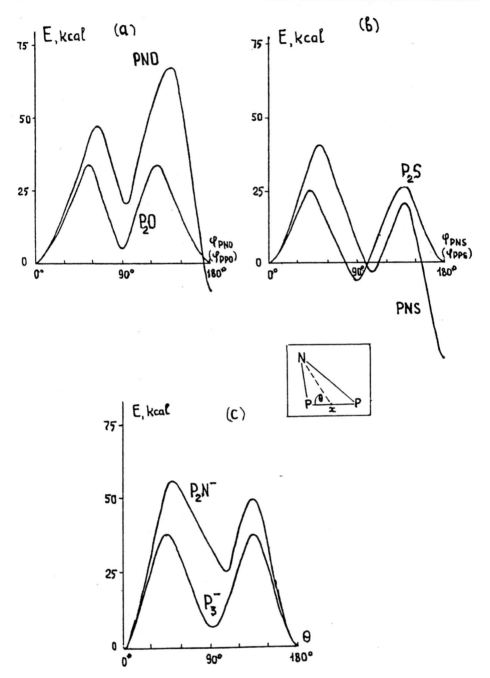

Figure 11. Profiles of PES along the MEP of the BAC ↔ ACB rearrangement in the systems PNO and P_2O (a); PNS and P_2S (b), N_3^-, NP_2^- and P_3^- (c) [161,162].

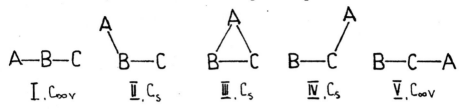

Figure 12. Alternative structures of the N_2O, NPO, P_2O and similar molecules.

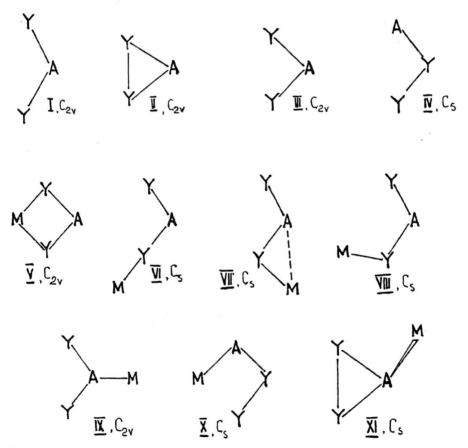

Figure 13. Alternative structures of triatomic AY_2 ligands with 18 and 20 valent electrons (I—IV), and their complexes MAY_2 (V—XI).

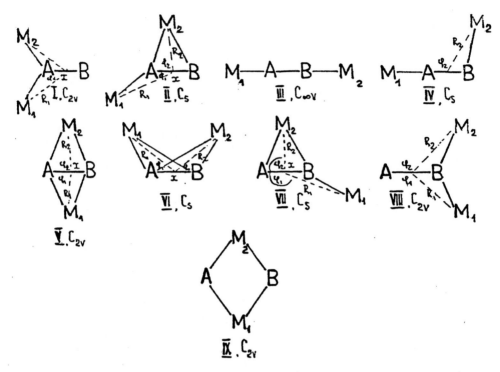

Figure 14. Alterantie structures of the M_2AB^+ ions with two cations M^+.

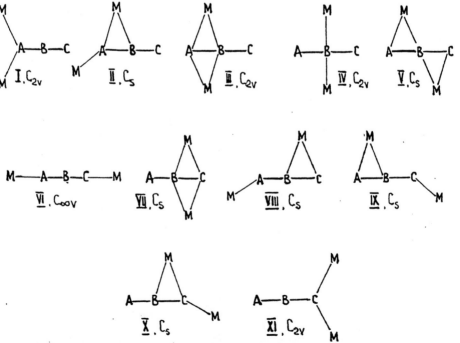

Figure 15. Alternative structures of the M_2ABC^+ ions.

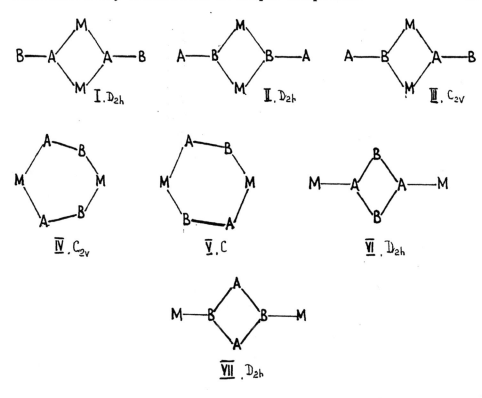

Figure 16. Alternative structures of the dimer $(MAB)_2$ molecules.

FERROCENE-BASED METALLOCHELATES

V.A. Kogan and A.N. Morozov

INTRODUCTION

The discovery of dicyclopentadienyliron, made independently by Pauson [1] and Miller [2] in 1951, has stimulated extensive studies concerned with the synthesis of new cyclopentadienyl complexes of both transition and non-transition metals. Later, these were included in a special class of metalloorganic compounds called p-complexes, together with the other coordination compounds containing various organic molecules with multiple bonds as the ligands.

Dicyclopentadienyliron, later named ferrocene, is a most widely studied sandwich type compound [3,4], not only due to its availability, but due to its possible application in various fields of industry as well [5-7].

From both practical and theoretical points of view, the derivative ferrocene-based molecular complexes which allow the obtaining of intra-complex compounds representing a series of heterometallic derivatives are the most interesting. These are applied, together with metallocene-containing polymers, as semiconductor materials in electronics [8], as catalysts in organic synthesis, etc. [9,10]. The complexes containing several heterometals within one molecule and behaving in some cases as the exchange clusters represent an interesting model system for the theory of exchange interactions.

There is some evidence that the transition metal chelate complexes containing the ferrocene fragment may act as the catalysts of styrene polymerization [11], similar to chelates based on the other p=complex, cyclopentadienylmanganese tricarbonyl (CMT, or cymantrene).

These compounds are also expected to exhibit an increased biological activity [12, 13] which suggests their possible use as the medicine. In this respect, especially promising are the complexes containing, besides the ferrocene, some other groups, e.g., hydrazones, whose biological activity is well established.

Investigation in the properties of ferrocene and the related compounds is stimulated not only by their possible applications, but by a permanent theoretical interest as well. While the ferrocene itself is sufficiently studied, its derivatives attract increasing attention. The main problem is the elucidation of mutual influence between the ferrocenyl fragment and the organic groups chemically bound to it in a molecule studied. One of the first principal results obtained was in establishing the fact that the closing of cyclopentadienyl rings in ferrocenophanes by various organic and elementoorganic groups allowed an essential and diverse modification of the molecular geometry [14] and properties [15]. Convincing evidence for this is found in an interesting observation that one of the cyclophane ferrocene derivatives is a potential catalyst of the solar light induced water decomposition [16], with a possible application in future energetics. However, until quite recently only some fragmentary works have been undertaken rather than a purposeful investigation in the modification of ferrocene fragment by complex formation. In our opinion, this problem is now a most intriguing one in both theoretical and practical aspects of ferrocene chemistry.

This review is one of the first attempts in generalizing the results of works concerned with various types of ferrocene metallochelates. Special attention is paid to the study of modification of ferrocene-containing ligands through complex formation, including the authors' original results (see Chapter 4).

1. Transition Metal Complexes With Ferrocene-Diketones

The chelating ferrocene ligands represent a sandwich fragment connected with one or several chelatophoric groups. Historically, the first such group was represented by a simplest b-diketone fragment [8, 10, 17-31]:

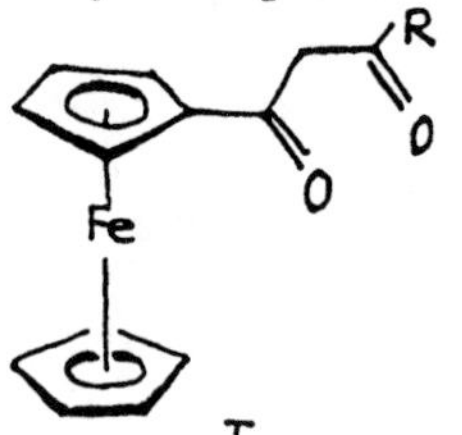

$R = CH_3, CF_3, C(CH_3)_3, Ph, Ph$

A high complex-forming ability of this ligand allowed the obtaining of chelate complexes with almost all transition metals of the periodic table. A comparative spectroscopic study (IR, UV, and ^1H NMR) suggests the following structure of complexes based on the type I ligand:

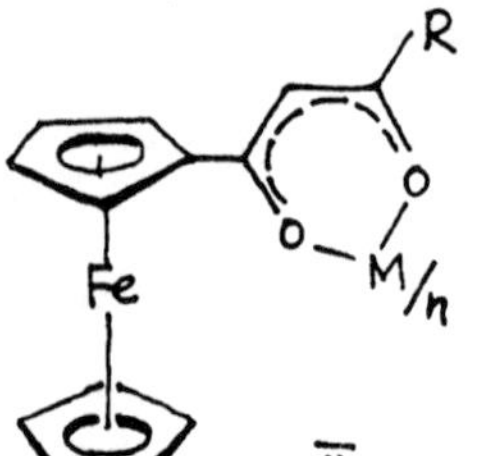

$M = Cu^{+2}, Th^{+4}, Pb^{+2}, UO_2^{2+}, Co^{+2}, Ni^{+2},$

$Fe^{+2}, Ru^{+2} Mn^{+2}, Zn^{+2}, Cr^{+3}, Fe^{+3},$

$Y^{+3}, Nd^{+3}, Sm^{+3}, Eu^{+3}, Gd^{+3}, Dy^{+3},$

$Ho^{+3}, Er^{+3}, Tm^{+3}, Yb^{+3}, Ln^{+3};$

$n = 2, 3, 4;$

which fully coincides with the chelate cycle structure of the similar transition metal complexes with b-diketones, not containing the ferrocene fragment [32-35]. The IR spectra of ligand I ($R=CH^3$) and the type II complexes ($M=Mn^{+2}$, Co^{+2}, Ni^{+2}, Cu^{+2}, Zn^{+2}, Cr^{+3}, Fe^{+3}; n=2, 3 [26]) contain the absorption bands at 1550 and 1620 cm^{-1} (ferrocene b-diketone) shifting to 1510-1535 and 1540-1570 cm^{-1} upon complex formation, assigned to the stretching vibrations of the C=C and C=O bonds, respectively.

The electronic absorption spectrum of the initial ligand is characterized by the presence of two absorption bands at 305 and 468 nm. On the complex formation, the former band either exhibits a bathochromic shift to 323 nm ($M=Co^{+2}$), 319 nm (Ni^{+2}), 318 nm (Cu^{+2}), or remains at the same position (in complexes of divalent zinc and manganese and trivalent metals). The latter band shows a shift only in complexes of trivalent chromium (l_{max}=465 nm) and iron (l_{max}=440 nm). Such bands are obviously assigned to the d-d transitions in iron atom.

These data in combination with an elemental analysis showed the complexes studied to possess the chelate structure II. Similar results were also obtained for the complexes of other metals with the type I ligands studied by the same conventional physico-chemical methods. This revealed a full identity of these complexes and suggested that the coordination mode of ferrocene-containing b-diketones did not depend on the nature of both the complex-forming metal M and the substituent R.

Attempts to investigate the configuration of the chelate node in the above complexes have been undertaken [27, 31, 36, 37] and some structural features of these compounds have been elucidated. Thus, the divalent ions of manganese, cobalt, zinc, and nickel were found to form the $ML_2 \lozenge H_2O$ type complexes with aceto-acetylferrocene whose solubility dropped in a series $CHCl_3 > CH_2Cl_2 > C_6H_5Cl > C_6H_4(CH_3)_2 > C_6H_6$ with a decreasing polarity of the solvent [27]. All these complexes readily formed the adducts with pyridine, (also including the copper chelate), of the type of $ML_2 \lozenge 2Py$, except for the nickel chelate which produced the adduct with a composition $NiL_2 \lozenge Py \lozenge H_2O$. Magnetochemical, IR and electron spectroscopic data suggested an octahedral chelate configuration in bisaquacomplexes of cobalt (II), nickel (II), manganese (II), (and their pyridine adducts), and a planar-square configuration, in CuL_2 and ZnL_2 complexes.

Such a behavior is characteristic of the transition metal b-diketonates [38], and the ferrocene-containing complexes have no differences in this respect from the related compounds without metallocene substituents. However, the introduction of ferrocene in a complex also results in the appearance of some properties not typical of the ordinary b-diketonate complexes.

Thus, the cryometry data [27] show acetoacetylferrocene based bisaqua-complexes to be monomeric, except for $NiL_2 \lozenge 2H_2O$ whose molecular weight depends on the solvent polarity and corresponds to a monomeric chelate structure in polar solvents, ($CHBr_3$(0.99 D) and $C_6H_5NO_2$ (3.99 D)), and to a dimeric structure in a low-polarity solvent (dioxane (0.45 D)). However, this observation contradicts to the results of Reference 31, where the cryometry data suggest a monomeric $NiL_2 \lozenge 2H_2O$ complex to be formed in both dioxane and chloroform.

The results reported in References 27 and 31 are also at variance as concerns the composition of the pyridine addict of this chelate. While a $NiL_2 \lozenge 2Py$ structure has been proposed in Reference 31, the authors of Reference 27 believe that only one water molecule of the aquacomplex is substituted by pyridine. The authors of Reference 31 noted an unusual behavior of the dehydrated nickel (II) complex whose effective magnetic moment in chloroform increased with time from 0.76 m_B up to 1.60 m_B*. This fact was explained by the existence of two equilibrium forms of the complex in solution, with a diamagnetic square-planar and a paramagnetic tetrahedral configurations of the chelate node. The effective magnetic moments of the complex measured in various solvents showed either zero or the "usual" values typical of a tetrahedrally coordinated central ion. This example demonstrates that the introduction of a ferrocenyl fragment in bis-chelates of transition metals containing the b-diketone type organic ligands may affect the properties of the chelate node itself due to an increase in the complex liability.

As for the complexes of divalent copper, manganese and cobalt with 1-ferrocenyl-1,3-pentadione and their dihydrates, References 27 and 31 agree in establishing their monomeric structures.

This fact contradicts a commonly accepted [38] tendency of b-diketonates to polymerization. The discrepancy is especially dramatic for the complex of nickel (II), since the trimeric structure of nickel (II) bis (pentane-2, 4-dionate) is well known [39, 40]. In References 27 and 31, the anomalous behavior of the chelates studied was explained by the inhibition of complex trimerization due to an increased steric interaction between the voluminous ferrocenyl substituents.

The electronic spectra of complexes exhibited the bands with absorption maxima at 38000, 32500-29500, and 22000 cm^{-1}, assigned to the electron transitions in a cyclopentadienyl ring and in a hydrogen-bound chelate cycle (p-p*-transition), and to a charge-transfer transition between the cyclopentadienyl ligand and an iron ore, respectively.

It has been noted that the IR absorption band at 3400 cm^{-1}, not observed in dehydrated metallochelates, seems to be due to the stretching OH vibrations of the coordinated water molecules. The high-frequency shift in the M-O stretching vibrations in the $MnL_2 < CoL_2 < NiL_2 < CuL_2$ (420-428 cm^{-1}) order suggests an increase in the complex stability in this sequence, whereby the less stable cobalt chelate is capable of oxidation in solution.

There have been attempts to study the effects produced by the electron-donor ferrocene fragment on the chelate cycle and on general properties of ferrocene-containing ligands and complexes [30]. However, it should be noted that the problem of such an influence is not always unambiguously solved.

Thus, the ESR study of b-diketonates of nickel (II), copper (II), cobalt (II), manganese (II), chromium (II), and vanadyl (II), either containing or not

---

* The absorption spectrum of this chelate in chloroform also exhibited a time dependence.

containing the ferrocene fragments, showed no evidence for the presence of such influence. The parameters of ESR spectra of the compounds obtained were the same as those of the similar b-diketonates not containing a sandwich fragment in their structure, this suggested a resemblance to the structures of chelate nodes in these compounds with a C_{4v}-symmetry surrounding of the central ion. A comparison of displacement coefficients calculated for the MOs of unpaired electrons in these complexes showed the values for $d_{x^2-y^2}$ orbitals to coincide within a ±0.032% accuracy. This is an unambiguous evidence for the same spin distribution on the b_1-orbital of both compounds and, hence, for the absence of any noticeable influence of this fragment on this orbital.

The X-ray diffraction patterns observed on the powdered samples of these chelates show the identical sets of diffraction maxima, thus indicating the resemblance not only of the molecular, but also of the crystal structures.

In our opinion, this last observation seems to be rather unexpected because a much more voluminous ferrocenyl radical, (as compared to methyl), must play a more essential role in packing the crystals of complexes in which this fragment is contained.

The b-diketonates of transition metals are known to initiate the polymerization and copolimerization of compounds with double bonds [11, 41]. Of special interest in this respect are the various heterometallic derivatives, which have stimulated an attempt [10] to study the effects of ferrocene, (and cymantrene), substituent on the catalytic and initiating properties of b-diketone complexes of some metals (Mn^{+2}, Pb^{+2}, Ni^{+2}, Cr^{+2}, Fe^{+3}, Eu^{+3}, Nd^{+3}, Sm^{+3}).

A derivatographic investigation of the thermostability of complexes revealed a higher stability of the ferrocene-containing complexes (up to 300°C) as compared to the similar CMT complexes (stable up to 230°C) and a higher initiating ability of the former styrene polymerization which was especially pronounced in iron (III) and nickel (II) compounds.

In Reference 10, a mixed chromium salt-chelate of ferrocene b-diketone and CMT (III) was obtained via the following reaction:

This compound represents an efficient antistatic addition in fuels which increases their electrical conductivity at zero electrizations [10].

There are data on an interesting series of ferrocene-containing semichelates in which the b-diketones fragment of ferrocene ligand forms a complex with some metal connected by a multi-center p-bond with other heteroligands [23, 42-44]. Examples are given by the compounds (IV) to (VI).

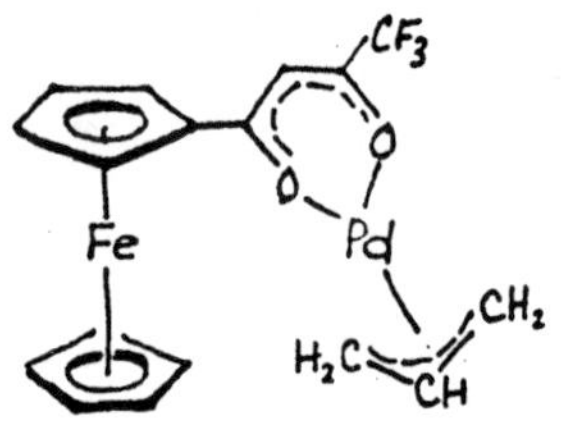

AllPd(FTA), IV

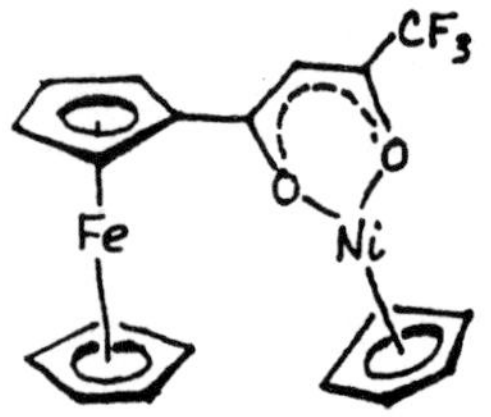

CpNi(FTA), V

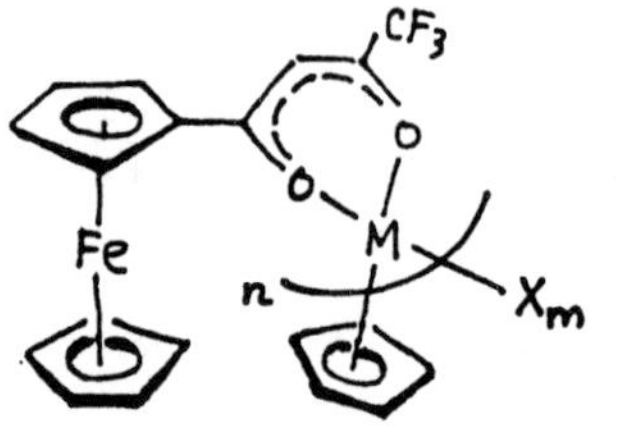

CpMX$_m$(FTA)$_n$, VI;
for M=Ti, Zr, Hf, Nb;
n=2, m=1, X=Cl$^-$

The presence, on the one hand, of ferrocene fragments and, on the other hand, of trifluoromethyl radical in the above structures is the main factor determining a much greater stability of these complexes as compared to their analogs not containing at least one of the substituents indicated. Thus, the semi-chelate IV is considerably more stable than the other b-diketonates of allyl-palladium and decomposes only at 135°C [42], while the latter compounds are instable already at room temperature.

It appears that the role of fluorine consists in increasing the coupling of the b-diketone group both by itself and with the p-system of the cyclopentadiene ligand in ferrocene upon coordination, which significantly stabilizes the complex.

A further insight in this problem is provided by a detailed investigation of the IR spectra of such compounds as cyclopentadienylnickel (V) trifluoroacetylferrocenoylmethanate and the initial ferrocenoyltrifluoracetylmethane (NFTA) and cyclopentadienyl-dicarbonylnickel.

The presence of unchanged bands belonging to the chelate cycle, metal-containing fragments, ferrocenyl and trifluoromethyl groups allowed the authors of Reference 44 to conclude that the nature of the M-(h^5-C$_5$H$_5$) bond is retained on the passage from pure metallocene to semi-chelates, in spite of some weakening of the bond. The equivalence of all C-C and C-H bonds in the Cp-ring is also not violated.

The role of ferrocene substituent is, apparently, to increase the interaction of metal with the Cp-ligand acting as a donor of electrons. The ferrocene itself exhibits, according to the authors of Reference 44, strong acceptor properties.

The donor ability of ferrocene should also be mentioned, so that the ligand systems considered are, in principle, the bifunctional ones (and even the polyfunctional due to a donor abilities of cyclopentadiene ligands [45-1]. It is important that a further increase in the polydentate character of ligands can be provided by the introduction of additional donor centers in the ferrocene ligand molecule. An approach to such a modification, resulting in that a ferrocene-containing ligand acquires one or several additional groups with heteroatoms, has been effected by substituting the neutral radical R in the b-diketone fragment (I) by pyridyl [52]. The obtained 2-, 3-, or 4-pyridoylferrocenoylmethane (VII-IX) is a polyfunctional donor:

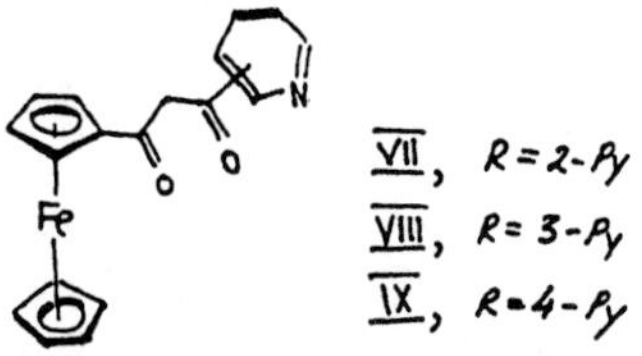

VII, R=2-Py
VIII, R=3-Py
IX, R=4-Py

The most studied 2-pyridoylferrocenoylmethane VII occurs in a solution in three tautomeric forms, the b-diketone VII and the enolized structures VII[a] and VII[b] with differently localized hydrogen bonds:

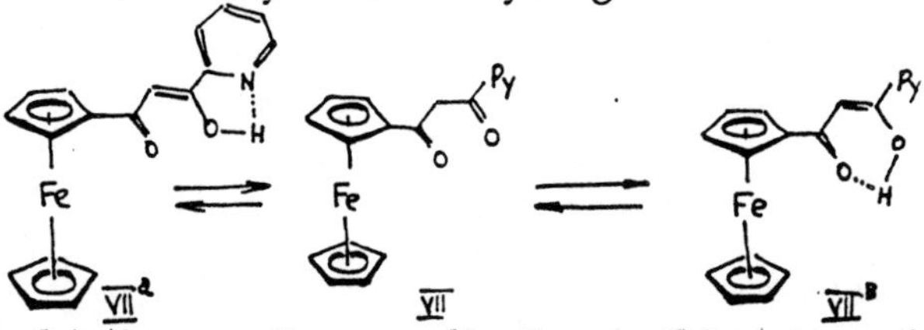

The study of competing coordination in this system during the complex formation with divalent copper showed that in the case of chelate the typical VII[b] form (II, R=Py, M=Cu[+2]) is produced. The pyridine nitrogen atom remains non-coordinated, while in a solid state the ligand is proposed to have the structure VII[a]. However, the pyridine nitrogen atom already in the copper chelate, i.e. upon the action of $CuCl_2$, exhibits more or less a pronounced trend towards coordination, (depending on the pH of medium), and forms several stable adducts with the Fe/Cu atomic ratio from 1:0.6 to 1:3. Some of these complexes have been isolated in a free form, with a supposed structure of the following:

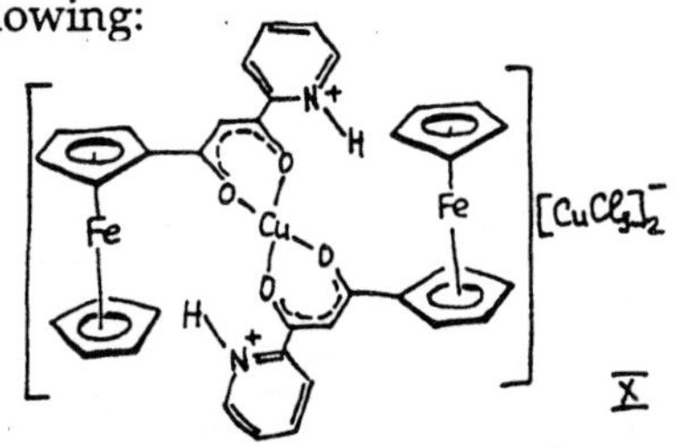

which agrees with the tendency of ligand VII to occur in the form of VII[a] containing a protonated nitrogen atom in the heterocycle.

Although the authors of Reference 52 deny the possibility of oxidation of the ferrocene structure to the ferricinium ion, this seems to be a quite reasonable situation [53] which is confirmed, in the authors' opinion, by the blue color of solutions of the compound X. Such color is typical of the salts of this cation [54]. It is possible that there is an equilibrium in the system which shifts to the left if one tries to isolate the complex X in a solid state or to extract it using an aprotic solvent. (Benzene allowed the extraction of only the red modification from blue solutions of X; probably, the extract contained only divalent iron [52]). In the authors opinion, it was a premature conclusion of Reference 52 that the pyridine nitrogen atom was non-competitive with the carbonyl oxygen atom, because no attempts were made to vary the Pirson acids according to principles of the theory of hard and soft acids and bases [55-57]. It must also be taken into account that the mode of localization of coordination bonds in polyfunctional donor systems, (represented by complexes VII), may be significantly affected by external factors, that is, by the reaction conditions [58, 59].

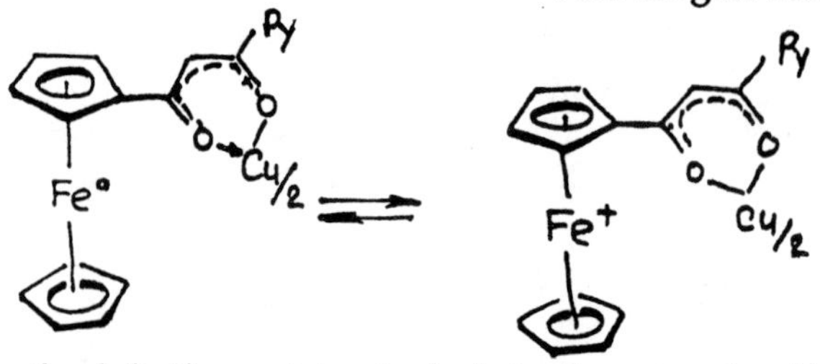

An increase in the polydentate character of the ferrocene based ligand systems through the introduction of a second b-diketonate group leads to the appearance of complex structures which are very interesting from a theoretical viewpoint.

For example, the complex formation reactions of uranyl (II) with mono- and bis-b-diketones of ferrocene of the I and XI types have been studied in Reference 36.

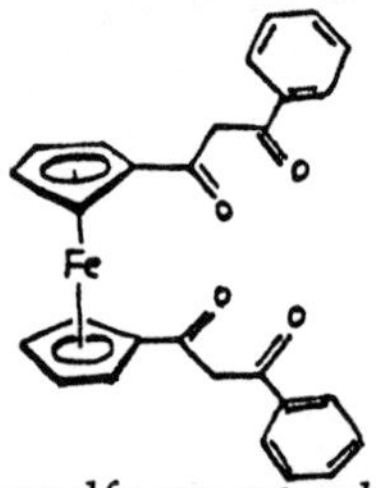

XI, H_2L

The uranyl bis-benzoylferrocenoylmethanate (XII, M=UO_2 (II)) has been synthesized in ethanol and its composition is described by the formula $UO_2L_2\lozenge(H_2O)$. In this compound, the equatorial coordination number of uranyl upon the complex formation reaches 5. It has been noted that water molecules are readily substituted in benzene solutions by the different monodentate ligands, such as Bu_3PO, pyridine, and DMSO. On the basis of IR spectroscopy data, the authors of Reference 36 suggested these complexes to possess the following structure type:

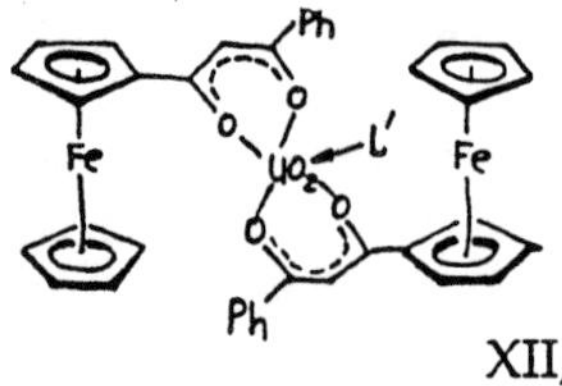

XII, L'=H_2O, Bu_3PO, DMSO, Py.

However, further investigation allowed a refinement of the above structure which proved to have a dimeric nature [37]. This was revealed by the X-ray diffraction showing the uranyl (II) complexes with ferrocenoylbenzoylmethane to represent the following configuration

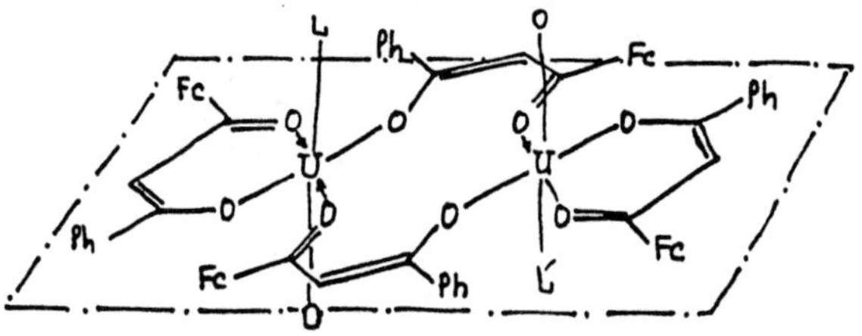

XIII, $(UO_2)_2L_4L'_2$; Fc=ferrocenyl
L' is a molecule of neutral monodentate ligand
(Me_2CO, Ph_3PO, Ph_3AsO, Me_2SO, Py, PyNÆO)

In this structure, the coordination number of uranyl increases to 6 for $L'=H_2O$. A comparative spectroscopic investigation of the obtained complexes and the ligand suggested a lack of the direct contact between the U^{+4} and Fe^{+2} ions, as revealed by a very low difference of the parameters of electronic spectra of the compounds studied.

During the investigation of the complex formation of bis-1, 1'-di (phenyl-propane-1,3-dione)-ferrocene with uranyl (II), the formation of either monomeric (XIV) or polymeric (XV) chelate structures was proposed. However, the measurement of molecular weights of the complexes by the Rust method allowed the monomeric configuration XIV to be excluded from consideration. As a result, the uranyl (II) complex with bis-1, 1'-di(phenyl-propane-1, 3-dione) ferrocene was assigned the polymeric structure XV.

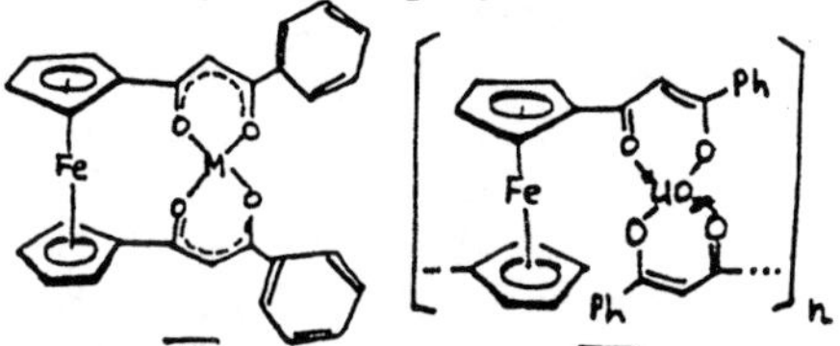

XIV   XV

The authors of Reference 36 observed a bathochromic shift by 10 and 20 nm of the absorption bands with l_{max}=340 and 422 nm in the electronic spectra of ligands in $UO_2L_2\Diamond H_2O$ and $UO_2L\Diamond H_2O_n$, respectively. It seems tat these bands can be assigned to the d-d transitions in iron ion and characterize the retaining the ferrocene structure in the XIII and XV complexes.

At the same time, it has been found that this ferrocene bis-b-diketone ligand XI and the similar compounds are capable of forming monomeric metal chelates [60-63].

On the basis of IR spectroscopy data in combination with elemental analysis and the cryoscopic molecular weight determination, the copper chelates were assigned the monomeric structures II and XIV. The reaction of these complexes with diluted phosphoric acid leads to the formation of the initial ferrocene bis-b-diketone. It has been pointed out that the transformation of the bis-chelating ligand X to the monomeric chelate XIV does not contradict the fact that the ligand system XVI presented below forms only the polymeric metallochelates [64]. According to Reference 60, the cis-positions of the two b-diketone groups in the chelate XIV must not be taken as a direct evidence of their cis-positions in the initial structure XI because the b-diketone groups may turn to this position under definite conditions [65] due to a low barrier for the rotation about the main axis of the ferrocene fragment. The initial ligand may be proposed to have the trans-conformation [66], because the X-ray diffraction data shows the acetyl groups of the two rings of 1,1'-diacetylferrocene to form an angle of about 144° (closer to their trans-position). This implies that the acetyl groups occupy positions 1 and 3' in ferrocene whose conformation is close to the prismatic (shadowed) XI[a]. However, this cannot prove the existence of ferrocene bis-b-diketone with the same configuration because of a growing probability of the cis-configuration XVII due to its stabilization upon the formation of an intramolecular hydrogen bond.

The monomeric configuration XIV appears to be sterically stressed. In this connection, important information was gained from a detailed comparative IR spectroscopic study of the I, XI, XVI ligands, and their copper complexes. The results obtained showed only a partial equivalence of complexes XIV to the rest of chelates because the IR spectrum of the stressed configuration XIV exhibited the absorption band at 1714 cm^{-1} not observed in the spectra of other complexes with similar structures. It has been suggested that this band appears within the absorption range of a non-conjugated ketone group due to a somewhat "incomplete" bonding of this group resulting from steric constraints involved in the coordination of the metal heteroatom by oxygen atoms.

Besides this, the IR spectra of all complexes contain the absorption band assigned to the coordinated C=O group (which excludes the formation of salts). The above considerations allow the XIV[a] structure to be represented as follows:

It should be noted that the authors of Reference 60, when discussing the monomeric structure XIV, have missed an important detail. Obviously, the structural stresses arising upon the formation of XIV can be essentially removed not only by the reduction in the chelate node symmetry, but by distorting the ferrocene fragment as well. It is believed that a wedge-shaped distortion [14, 67, 68] of the fragment renders the "distant" oxygen atoms of the chelate cycle closer and strengthens (shortens) the C-O and M-O=C bonds. For these reasons, the proposed structure XIV^b appears to us as a more adequate than the above configuration XIV^a:

The results of References 37 and 60 are not contradictory, but rather complementary and, if treated in combination, allow an important conclusion that the formation of alternative poly- or monomeric structures during the metal complex formation with metallocene-containing ligands having the chelatophoric substituents in both rings is significantly affected by the dimensions of both the complex-forming metal itself and the chelate cycle.

From this point of view, the polymeric structure of the complex XV is determined, (in contrast to the configuration XIV^b), by large distortions of the ferrocene fragment and chelate node due to a high ionic radius of UO_2^{+2} as compared to Cu^{+2}.

It is interesting to note that the other metals, (divalent iron, cobalt, nickel, manganese, and vanadyl), also form the polymeric XI type chelates with the $ML\lozenge2H_2O$ composition [61-63], while the monomeric XIV configuration (for $M=Cu^{+2}$) has once again been confirmed by the data of IR, electron, and gamma-resonance spectroscopies [61]. A higher tendency of this complex to form adducts with organic bases (together with some features of the ESR spectrum of XIV) has been attributed to stresses in the chelate cycle arising due to the steric constraints of the structure.

In addition to the b-diketone substituents in the rings, the chelating groups are often represented by the hydrazine derivatives, e.g., by hydrazone, thiosemicarbazone, and the other similar groups.

2. Metallochelates Based on Thiosemicarbazones and Hydrazones of Acyl Ferrocene Derivatives

A distinctive feature of organic derivatives, acid hydrazides and hydrazones, is their high complex forming ability. Various problems concerning the structure, properties, and applications of the transition metal complexes with hydrazones of various aldehydes and ketones were treated in numerous works reviewed in References 69 to 72. At the same time, the authors note that the studies of the complex-forming properties of the hydrazine derivatives containing the ferrocene fragment are few and scattered [73-80]. A general formula of ligands of this type described in literature can be represented as:

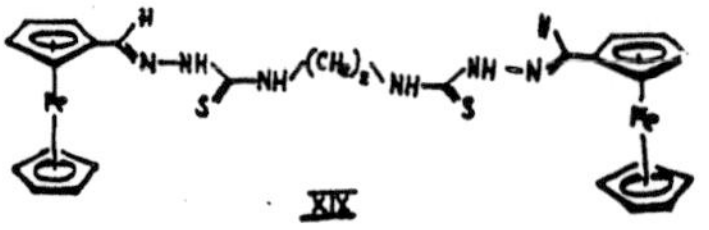

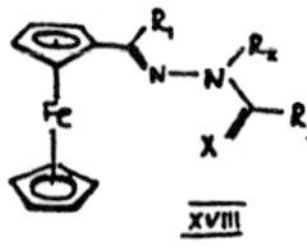

$X=O$, S; $R_1=H$, CH_3; $R_2=H$, CH_3; $R_3=NR'R''$, Ar, SCH_3, Py; $R'=H$, CH_3; $R''=H$, Ph, $(CH_2)Ph$, $3NO_2Ph$, $4NO_2Ph$, C_2H_5, $CH_2=CHCH_2$, $CH_3(CH_2)_6$

It has been reported [73] that no complexes with the configuration XVIII are formed with $R_2=CH_3$, this shows evidence for the preferential coordination of ligands in the deprotonated form only.

Taking the co-existence of the equilibrium keto-imine and enimine forms of XVIII in solutions of these ligand systems into account,

the authors set the problem of coordination of various tautomeric forms of hydrazone, (or the related compounds), during the complex formation. In most cases [73-80], this problem is uniquely solved for the enolization of hydrazones upon the complex formation with the transition metal ions (except for the platinum (II)). This has been confirmed by the data of IR, ESR, and PMR spectroscopies which allowed the transition metal complexes with ferrocene-containing derivatives of the hydrazine XVIII ($R_2=Me$) to be assigned the structure XX:

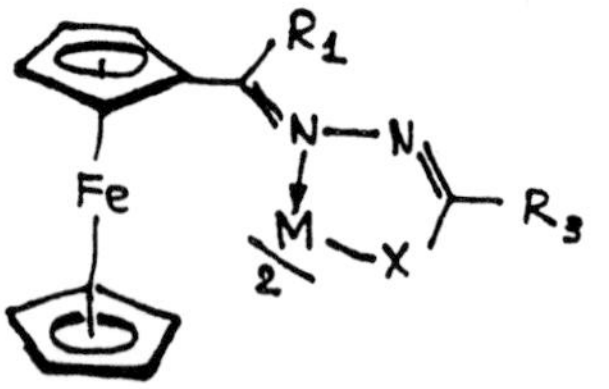

XX, $M=Cu^{+2}$, Ni^{+2}

The typical changes observed in the IR spectra of complexes XX as compared to those of the ligands XVIII consist in the vanishing of the n(N-H) and n(C=O) absorption bands with a simultaneous shift of the n(N-N) band to the short wavelength range. Evidence for the chelate structure of these complexes is also presented by the appearance of new IR absorption bands at 1280-1220 cm^{-1} and 1620-1600 cm^{-1}, assigned to the n(C-O) and the vibrations of conjugated hydrazone group , respectively. In addition, the IR spectra of the complexes studied contain practically unchanged absorption bands characterizing a stable ferrocene structure.

The coordination via the enole oxygen atom and the hydrazone nitrogen atom (corresponding to the configuration XX), which is a typical mode for aroylhydrazones [81-83], was unambiguously identified using the X-ray diffraction data for the dioxomolybdenum (VI) complex with monoacetylferrocene benzoylhydrazone having a transstructure. In these chelate molecules, the cyclopentadienyl rings are acoplanar to each other and to phenyl substituents in the hydrazone group [80].

The bathochromic shift observed for the band with the absorption maximum at $l_{max}=455$ nm in the electronic spectra of complexes, accompanied by a simultaneous increase in the band intensity, was attributed to a growth in the p-coupling of cyclopentadiene rings and the hydrazone group [80]. In the authors opinion, this interpretation is not completely consistent with the known angle (12.5°) between the planes of cyclopentadiene rings and the chelate cycle. Although not too large, the angle is quite sufficient for the violation of such a coupling, the more that this band refers to the d-d transition in the iron ion [84].

The question of whether the electron-donor ferrocene fragment affects the properties of complexes and ligands and which it is contained still remains open. The electron diffraction study of both the complex and ligand considered gave a negative answer [80], in contrast, e.g., to References 42-44, 25 or References 74-75.

The electron spectroscopic, polarographic, and amperometric data allowed the authors of References 74 and 75 to conclude that the ferrocene substituent exhibited a greater effect on the properties of a metal chelate as compared to the ligand (in the process $Fe(C_5H_5)_2)-\bar{e}ÆFe(C_5H_5)_2^+$). It was noted that the obtained complexes contained three electrochemically active centers, thiosemicarbazone group and Fe^{2+} and M^{2+} ions (the latter ion was a probable competitor to the ferrous ion during its reversible oxidation).

The authors of Reference 76 reported on the investigation of complex-forming properties of 1,1'-diacetylferrocene bis-thiosemicarbazone. This compound (XXI) formed complexes with copper (II) having a proposed polymeric structure XXII:

XXI   XXII

Unfortunately, no data were presented to justify the metal chelate structure proposed, although some features in the IR spectra showed evidence for the coordination of a thiol ligand form by metal atoms. However, the data [85-90] suggests that the ferrocene-containing hydrazones are not always coordinated in this way, that is, via the enole oxygen atom (thiol sulfur atom) and the nitrogen atom of the hydrazone fragment [81-83]. An exception to this rule has been observed for the divalent palladium complexes, apparently, due to the tendency of entering the orthocyclo-palladate formation reactions. This allows for the obtaining of the following types of compounds:

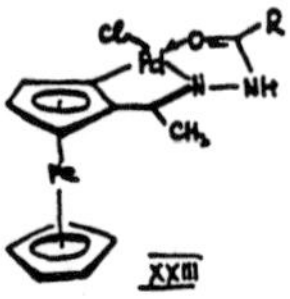

As can be seen, the oxygen atom in these complexes is coordinated within the ketone group, in agreement with the observed shift of n(C=O) in the IR spectra of complexes by 90 cm^{-1} to longer wavelengths as compared to the ligand spectrum, which is typical of the coordination of free ketone groups. This coordination mode of 1,1'-diacetylferrocene bis-acetylhydrazone upon the Pd^{2+} complex formation is also substantiated by the PMR data showing the vanishing of the signal from one proton of each ring in the spectrum of complex, in contrast to that of ligand 90.

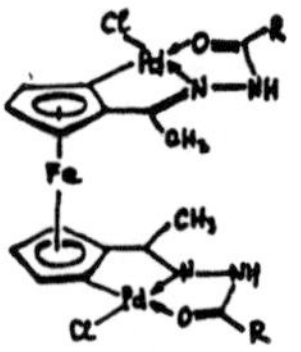

It has been noted that the XXIV type complexes are capable of substituting the coordinated ketone group of the hydrazone fragment by a stronger donor, triphenylphosphine, to form the configuration XXV:

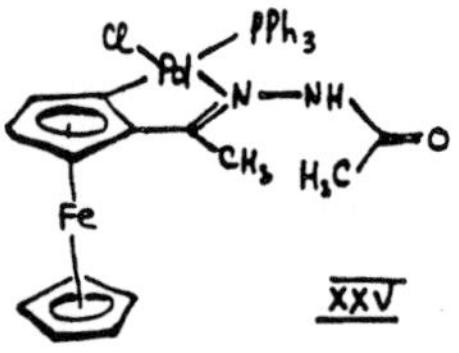

Finally, the authors would like to note the practical value of investigations concerning the ferrocene hydrazone derivatives. On the one hand, this interest as been stimulated by the biological activity of these compounds related to their ability of chelating the metal-containing centers in viruses, thus, increasing the resistance to such decreases as tuberculosis, etc. [93]. On the other hand, the introduction of the ferrocene fragment into penicillin has also been known to affect its biological activity [12, 13]. Naturally, some investigators have attempted to combine the two biologically active fragments.

3. Other Heterometallic
Ferrocene-Containing Metallochelates

The communications on ferrocene-containing complexes and ligands including the other chelatophoric groups besides the b-diketone and aroylhydra-
zone (thiosemicarbazone) are few and may appear to have a rather casual
character.

For example, such additional chelating ligands can be represented by the
condensed derivatives based on b-diketones of ferrocene and various aromatic
and aliphatic amines [94-97].

One of the first complexes with this type of ligand was a trinuclear
nickel chelate XXVI obtained by template synthesis [94] (en = ethylene diamine):

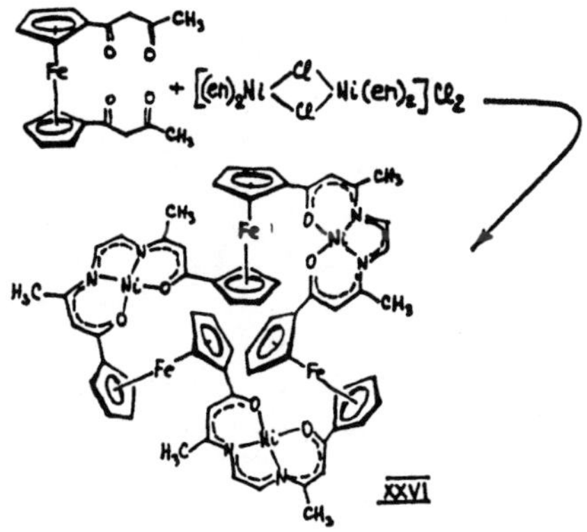

References 95 to 97 were concerned with the synthesis and investigation
of the polymeric chelates based on the condensation products of the above b-
dicarbonyl ferrocene compounds with ethylene diamine, m-phenylenedi-
amine, hexamethylenediamine, and benzidine with metals (Cu^{+2}, Mn^{+2},
Co^{+2}, Cr^{+3}, Fe^{+3}), taken in a five-fold excess for the obtaining of poly-
chelates. The data of IR spectroscopy and differential thermal analysis allowed the complexes obtained to be assigned polymeric structures. It was also
reported on the synthesis of ferrocene dianyles by condensation of ferrocene
bis-b-diketone with H_2H-R amines (R=C_4H_9, C_5H_{11}, C_6H_5 and $C_{10}H_7$; and
the bis-b-diketone: H_2N-R ratio of 1:3). The so obtained dianyles exhibited
the absorption bands of $n_{C=O}$ and $n_{C=N}$ in the IR spectra and formed poly-
chelate compounds with Cu^{+2}, Co^{+2}, Mn^{+2}, and Fe^{+3}. The conclusions of References 95 to 97 are rather qualitative, and it would be necessary to use a
more diverse set of analytical techniques in order to corroborate the polymeric state of the chelates synthesized.

Another type of ferrocene-containing chelating ligands seems to be represented by mono- and diferrocenecarboxylic acids [98] and their thio-derivatives [99, 100].

These ligands have been proven by the IR and electron spectroscopy data to form the complexes of the XXX type with the carbonyl group taking part in coordination:

$$X=O, S;$$
$$M=Cu^{+2}, Zn^{+2}$$

The ESR spectroscopy study of copper (II) complexes with ferrocene-carboxylic and ferrocenedithiocarboxylic acids [99] revealed a stronger rhombohedral distortion of the chelate node in the first of the above complexes. It was proposed that the ferrocene fragment plays a predominant role in determining the electronic structure of the whole complex.

An interesting and prospective direction of synthesis of novel chelating ferrocene ligands is related with the obtaining of the XXXI type Schiff bases [101] and their derivatives containing the oxime groups [102-103].

The first successful result of this work was the synthesis of the XXXII and XXXIII type compounds and the highly symmetric metallochelate XXXIV:

$$XXXIII,$$
$$M=Fe^{+2}, Ni^{+2}, Co^{+2}, Cu^{+2}, Mn^{+2}, Cr^{+3}$$

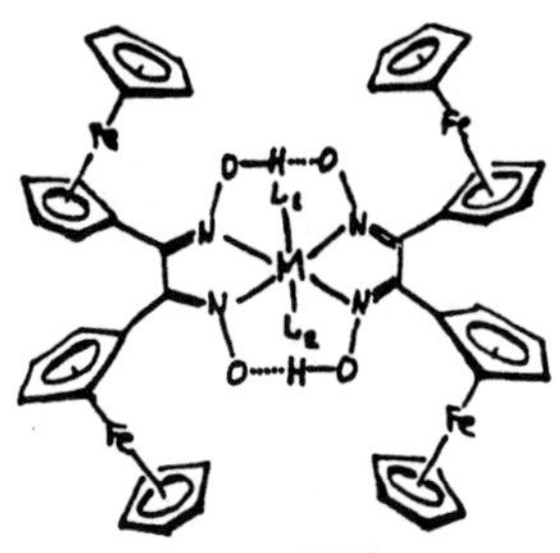

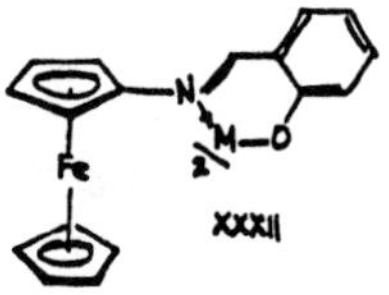

xxxii

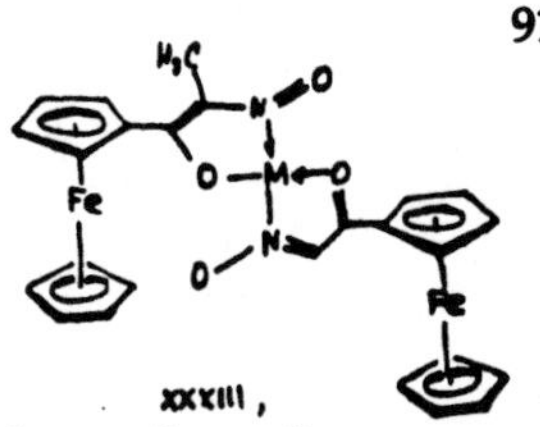

xxxiii,

M=Ni^{+2}, Pd^{+2}, Pt^{+2}, L$_2$=L$_1$=O
M=Pt^{+4}, L$_2$=L$_1$=Cl$^-$
M=Co^{+3}; L=Cl$^-$, L$_2$=Py

xxxiv

Our first communication on the synthesis of transition metal complexes with 1, 1'-diacetylferrocene bis-aroylhydrazones was followed by the other work [104[which described a similar ligand and its complexes. However, the polymeric structure of these complexes was not substantiated by the molecular weight measurements (Reference 104), and their chemical properties did not agree with the results presented in Reference 105-111.

It must be noted that the interest in the complexes with various ligands containing metallocene fragments is permanently growing and new important data providing a deeper insight in the nature of sandwich compounds have been obtained. At the same time, an analysis of the data has revealed that not enough attention has been paid to the problem of mode and degree of modification of metallocene-containing ligands during the complex formation. Some aspects of this problem are treated in the next chapter.

4. THE PROBLEM OF STRUCTURAL MODIFICATION OF FERROCENE BY THE METHOD OF CHELATE FORMATION

This chapter deals with a complex compounds of the type of XXXV containing ferrocene as a fragment of the metal-chelate cycle:

xxxv

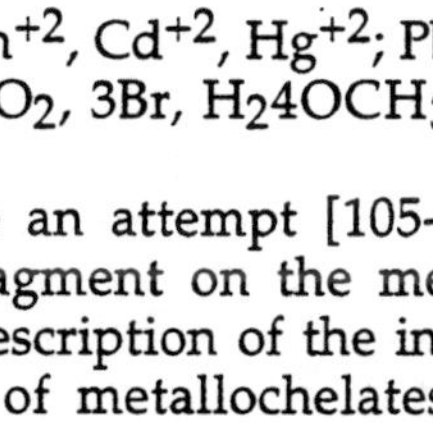

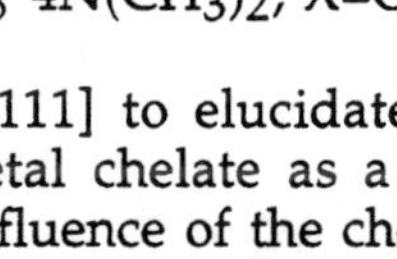

xxxvi

M$_1$=Fe^{+2}, Ru^{+2}; M$_2$=Zn^{+2}, Cd^{+2}, Hg^{+2}; Pb^{+2}, Pd^{+2}, Cu^{+2}, Co^{+2}, Ni^{+2}
R$_2$=NH$_2$; Ph-R, R=3NO$_2$, 3Br, H$_2$4OCH$_3$ 4N(CH$_3$)$_2$; X=O, S.

These works were an attempt [105-111] to elucidate the effects of ferrocene as a ligand fragment on the metal chelate as a whole, (theoretical backgrounds for the description of the influence of the chelate cycle composition on the structure of metallochelates were presented in References 112 and114) and, also, to study the possibility of controlling the ferrocene geometry via the complex formation. This last point is of special interest for the present review, arising from a comparison of the behavior of the XXXV type metallochelates with that of ferrocenophanes.

Metallocenophanes are known to represent the cyclic derivatives of sandwich complexes, where the cyclopentadienyl rings are linked with one or several bridges composed of n atoms or atomic groups:

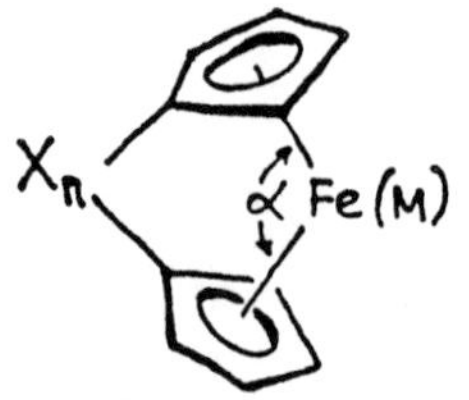

X-(CH$_2$)$_m$, P=Ph, Ge(Ph)$_2$, Si(Ph)$_2$, etc.; =152.3° for n=1[14]; 156.8°for n=2 [8]; and 171° for n=3 [114]. M=Fe, Ru,...

The steric stresses arising in these structures at low n values lead to the deformation of metallocene fragments whose extent depends on the length (n) of the bridging group. This is accompanied by acoplanarization of the cyclopentadienyl rings, whereby their tilt angle a decreases considerably [14, 67, 68].

The possibility of various modes of ferrocene fragment transformations is well known from the literature [14, 15, 114-116]. These distortions are represented schematically in Figure 1. It should be noted that several deformation types may occur simultaneously within one molecule, e.g., B+E [16], C+D+E [15], and in principle all of them at once.

In References 105 to 111, a coordination node of the type XXXV metallochelate linking the two cyclopentadienyl rings was used as the bridge (instead of ferrocenophanes), and a series of complex forming metals, (zinc (II), cadmium (II), mercury (II)), with a gradually increasing ionic radius was chosen. The metallochelates were synthesized by condensing diacetylferrocene with appropriate organic derivatives of hydrazine, and identified by elemented analysis and IR spectroscopy. The monomeric character of the complexes obtained was confirmed cyrometrically (by the Rust method).

The IR spectra of XXXVI ligands and XXXV complexes exhibited the absorption bands at 490 and 510 cm^{-1} corresponding to the rotation of cyclopentadienyl rings about the Fe-ring axis [78, 79]. The relative intensity of these bands falls in this order: ferrocene>ligand>chelate. This effect can be attributed to the limitations imposed on the rotation of cyclopentadienyl rings on their stabilization due to either the complex formation or the appearance of a stable intramolecular hydrogen bond in the ligand XXXVI:

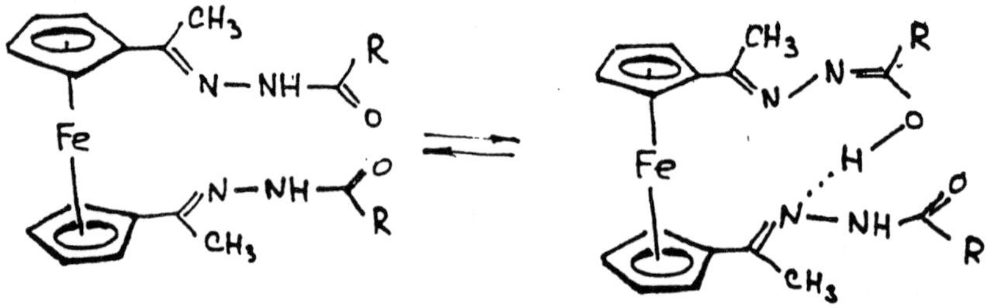

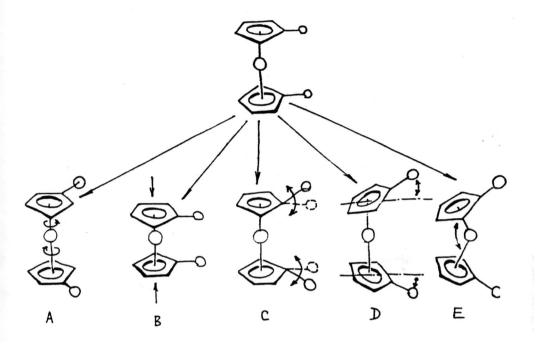

A   B   C   D   E

Figure 1. Possible types of sandwich deformations.

The possibility of such stabilization of cyclopentadienyl rings was proposed on the basis of theoretical considerations in Reference 115.

On the other hand, the acoplanarization of the sandwich fragment structure violates the equivalence of all the C=H and C=C bonds in the rings, which is manifested in both shapes and the positions of their absorption bands. Thus, the band of $n(C==C)$ at 1400 cm^{-1} splits into two components with the frequencies of 1402 and 1436 cm^{-1} upon a wedge-shaped distortion of the sandwich (ruthenocene) [117]. A similar effect has also been observed on the formation of the XXXV type chelates. For example, during the formation of a mercury (II) complex with 1,1'-diacetylferrocene bis-3-nitrophenylhydrazone, the $n(C==C)$ absorption band at 1384 cm^{-1} has been found to split into two components with the frequencies 1383 and 1368 cm^{-1}. The same effect is observed, e.g., in the XXXV type structures (M=Cu^{+2}, Ni^{+2}, Co^{+2}). Although the presence of other aromatic C==C bonds in the metallochelate structures may complicate a correct and unambiguous assignment of the absorption bands to the vibrations of long bonds, the application of this test [117] provides an indirect evidence for a wedge-shaped distortion of ferrocene fragment in a number of complexes.

Another method of revealing the wedge-shaped ferrocene distortion is provided by the electron spectroscopy and is based on the mobility of the absorption band at l_{max}=440 nm observed in the electronic spectra of ferrocene upon its wedge-shaped deformation [14]. A comparative study of the diffuse reflection spectra of the XXXV type complexes and XXXVI type ligands confirmed the existence of this spectral feature accompanying the above deformations in the compounds studied.

The parameters of diffuse reflection spectra of the ligand and the complexes of nickel, zinc, cadmium, and copper with 1,1'-diacetylferrocene bis-3-nitrobenzylhydrazone are given in Table 1.

Table 1. Parameters of diffuse reflection spectra of divalent nickel, zinc, cadmium, and mercury complexes with the XXXV type ligand (R=3-NO$_2$)

	l_{440} (nm)	l_{max} (nm)	n (cm^{-1})	n (cm^{-1})	E^{exp} (eV)	R_M (Å)
Ferrocene	450	–	22220			
H$_2$L (R=3-NO$_2$)	457	7	21850	370	0.046	0.69
NiL	472	22	21150	1070	0.134	0.69
XnL	473	23	21120	1100	0.137	0.74
CdL	484	34	20620	1600	0.200	0.97
HgL	495	45	20200	2020	0.253	1.10

During the investigation of the effect of ferrocene fragment distortions on the 440 nm absorption band shift, the authors of Reference 14 neglected the bathochromic shift of this band by 10 to 15 nm. Indeed, the maximum displacement of the absorption band at 440 nm may reach +45 nm [105] and is apparently related to a wedge-shaped deformation of the sandwich structure studied.

Another piece evidence justifying the use of the 440 nm band shift as a test for the wedge-shaped ferrocene fragment distortion is provided by the fact that such a shift is insignificant in the mon-3-nitrobenzoylhydrazone based complexes free of these distortions. Thus, in the mono-3-nitrobenzoyl-hydrazone complex of cadmium (II) the 440 nm band shift is as low as +10 nm, in contrast to the value of +34 nm observed in the cadmium (II) chelate based on 1,1'-diacetylferrocene bis-3-nitrobenzoylhydrazone.

The acoplanarization of cyclopentadienyl rings in the XXXV type complexes is also confirmed by a correlation of the above shift with the ionic radius of the complex forming metal, in agreement with the proposed enhancement of structural stresses in the XXXV type systems with an increase in the metal ion radius (Figure 2). Such a correlation can be explained by interpreting the mobility of the 440 nm band in the UV spectra of ferrocene using the diagrams of Lauher and Hoffmann [118]. The diagram shows a decrease in the energy gap between the boundary orbitals of metallocene upon the wedge-shaped deformation (depending on the tilt angle of the rings). In other words, it is indicative of a decrease in the energy which is necessary for the system to pass into an excited state, ($a_{1g}Æe_{1g}$; $2a_1Æ2a_1Æb_1$ in a wedge-shaped sandwich), this being just what implies a shift of the corresponding band in the diffuse reflection spectrum towards lower energies.

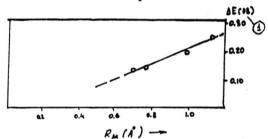

Figure 2. Variation of the d-d transition energy in ferrocene upon the chelate distortion vs. the radius of a complex-forming metal. Key: 1 - eV

Unfortunately, both the calculation itself and the diagram presented in Reference 118 reflect only some general regularities in the behavior of molecular orbitals of the sandwich during the acoplanarization of its structure, while providing no data which would allow a quantitative evaluation of the energy gap between the boundary orbitals for any particular metallocene, (bis-dicyclopentadienyliron), upon a distortion of its structure. A lack of these data still hinders the principally possible estimation of a decrease in the angle between the Fe-ring semiaxes as a function of the 440 nm band shift. However, an essential coincidence between the qualitative results of the authors, (and Reference 14 where the sandwich distortion in model compounds has been unambiguously confirmed by the X-ray diffraction data), and the calculations of Reference 118 seems to justify the use of the 440 nm band shift in the electronic spectra of the distorted systems studied as a test for the wedge-shaped sandwich deformation.

Positive results have also been obtained with the aid of the extended X-ray absorption fine structure (EXAFS) spectroscopy. Presently, this method is widely applied in the coordination chemistry [119, 120] and provides a highly accurate determination of the structure and geometrical parameters of the nearest coordination shell of a metal. The authors have used the EXAFS technique in order to confirm the presence of the ferrocene fragment distortion and to elucidate its character. The EXAFS spectra exhibited regular variations in the ferrocene order (diacetylferrocene) Æ ligand XXXVIÆcomplex XXXV ($M=Cd^{+2}$) which allowed the authors to make several conclusions. First, the ring vibrations are reduced due to a rigid linking of the monomeric structure; second, the ferrocene fragment itself is compressed, (and the separation between the cyclopentadienyl rings is decreased by 0.14 Å); and third, the EXAFS spectrum of the complex is sensitive to the second coordination shell of an iron atom due to a reduction of the Fe-Cd distance as a result of the wedge-shaped sandwich distortion. The EXAFS data obtained allows for the characterization of the total deformation of the sandwich structure of the XXXV type complexes as B+C+E (Figure 1) while the D type ring distortion is excluded.

Thus, a whole complex of physico-chemical studies suggests a possible wedge-shaped distortion of the ferrocene fragment in the XXXV type metallochelates. A direct experimental verification of this hypothesis is hindered by the impossibility of X-ray diffraction analysis, (due to insolubility of the chelates studied), and by the ambiguity in the interpretation of g-resonance spectroscopy data for the distorted ferrocenes [14, 15, 67, 68].

At the same time, the very structural distortions of the ferrocene fragment in the XXXV type complexes are responsible for the features in their chemical properties related to an increased accessibility of unshared electron pairs at the a_{1g} and e_{2g} levels of ferrocene in both the XXXV type chelates and the n-ferrocenophanes The data of References 105 to 107 showed evidence for the formation of stable paramagnetic adducts of ferrocene metallo chelates with various Lewis acids. These works established the paramagnetism of the XXXV type complexes and their ability to absorb oxygen and some other gases (Table 2).

Of special interest is the observed paramagnetism of the XXXV type complexes, since both ferrocene and the introduced complex-forming metal atoms ($M=Cd^{+2}$, Zn^{+2}, Hg^{+2}) are diamagnetic. Another important result is a linear dependence of the effective magnetic moments of the XXXV type complexes on the ionic radius of a complex-forming metal [122] (Figure 3).

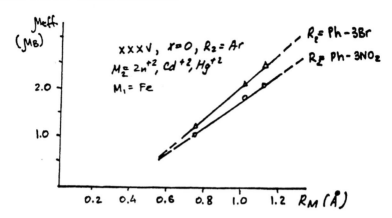

Figure 3. Effective magnetic moment of the XXXV type chelates vs. the radius of a complex-forming metal [122].

On the basis of a thorough and comprehensive analysis of the possible reasons for paramagnetism of the XXXV type ferrocene complexes with $M=Xn^{+2}$, Cd^{+2}, Hg^{+2}, etc., and taking the constancy of chemical shifts in the g-resonance spectra of the XXXVI type diamagnetic ligands into account, it has been concluded that the anomalous magnetic properties of these compounds are due to the molecular oxygen adsorbed during the synthesis. This proposition has been confirmed by the synthesis of the XXXV type compounds with $M=Xn^{+2}$, Cd^{+2}, Hg^{+2} in an argon atmosphere, whereby the obtained complexes have the same composition as above, but exhibit diamagnetic properties (Table 2).

It has been established that the complexes show a diamagnetic transition upon heating to 80° or under the action of various radiations (UV, X-ray, visible). The observed weight loss agrees quantitatively with the oxygen content in the complex calculated using the data of magnetochemical measurements.

The decriptometry and gas-chromatography data revealed the presence of molecular oxygen in the XXXV type complexes It was suggested that the compounds formed had a clathrate character, with oxygen included in the cavity $\underbrace{Fe-C_{cp}-C_{C=0}-N-M-N-C_{C=0}-C_{cp}}$ formed upon closing the chelate

cycle. It was also reported on the synthesis of ruthenocene based complexes ($M_1=Ru$; $X=S,O$) with a similar structure but a lower content of absorbed oxygen due to a greater size of the above cavity. This last circumstance reduces the steric obstacles preventing oxygen from leaving the complex in the case of a XXXV type structure ($M_1=Ru$; $M_2=Zn^{+2}$, Cd^{+2}, Hg^{+2}; $X=S$, O; $R=NH_2$, Ar).

Table 2. Data of magnetochemical measurements

Type of compound	R	R_I	M_I	M_2	T,K	meff. (m$_B$)	Synthesis conditions
XXXV	3-NO$_2$	CH$_3$	Fe	Zn	293.6	1.14	air
(R$_2$=Ar)	3-NO$_2$	CH$_3$	Fe	Zn	77.4	0.98	air
X=0	3-NO$_2$	CH$_3$	Fe	Zn	290.1	dia	argon
	3-NO$_2$	CH$_3$	Fe	Cd	293.6	1.83	air
	3-NO$_2$	CH$_3$	Fe	Cd	77.4	1.54	air
	3-NO$_2$	CH$_3$	Fe	Cd	289.1	dia	argon
	3-NO$_2$	CH$_3$	Fe	Hg	290.2	2.22	air
	3-NO$_2$	CH$_3$	Fe	Hg	77.4	1.51	air
	3-NO$_2$	CH$_3$	Fe	Hg	290.2	dia	argon
	3-NO$_2$	CH$_3$	Fe	Pd	293.1	1.18	air
	3-NO$_2$	CH$_3$	Fe	Pd	293.1	dia	argon
	3-NO$_2$	CH$_3$	Fe	Cd	293.1	2.29	air
	3-Br	CH$_3$	Fe	Zn	293.1	1.29	air
	3-Br	CH$_3$	Fe	Zn	293.1	dia	argon
	3-Br	CH$_3$	Fe	Hg	296.0	2.54	air .
	4-OCH$_3$	CH$_3$	Fe	Zn	293.1	dia	argon
	4-OCH$_3$	CH$_3$	Fe	Hg	293.1	dia	argon
	4-N(CH$_3$)$_2$	CH$_3$	Fe	Zn	295.3	1.11	air
	4-N(CH$_3$)$_2$	CH$_3$	Fe	Zn	294.0	dia	argon
	4-N(CH$_3$)$_2$	CH$_3$	Fe	Hg	293.1	dia	argon
	3-NO$_2$	CH$_3$	Ru	Cd	291.1	0.74	air
	3-NO$_2$	CH$_3$	Ru	Hg	292.1	0.80	air
	3-NO$_2$	CH	Ru	Hg	291.1	dia	argon
	3-Br	CH$_3$	Ru	Zn	292.1	dia	air
	3-Br	CH$_3$	Ru	Cd	292.1	1.06	air
XXXV	3-Br	CH$_3$	Ru	Cd	291.1	dia	argon
(R$_2$=Ar)	3-Br	CH$_3$	Ru	Hg	296.6	1.40	air
X=0	4-OCH$_3$	CH$_3$	Ru	Hg	291.8	dia	argon
	4-OCH$_3$	CH$_3$	Ru	Zn	293.3	dia	air
	–	CH$_3$	Ru	Cd	293.3	dia	air
	–	CH$_3$	Ru	Zn	291.4	dia	air
	–		Ru	Cd	291.4	dia	air
XXXV	–		Ru	Hg	289.3	1.16	air
(R$_2$=NH$_2$)	–		Ru	Hg	289.3	dia	argon
X=S							

Taking into account that the similar closed systems based on hydrazones and thiosemicarbazones (XXXVIII, XXXIX) contain no oxygen [123, 124], it has been concluded that an important role belongs to the iron atom of the ferrocene fragment whose oxygen affinity is well known [125, 126]. The possible mechanism of oxygen trapping is based on the effect of compression or "chelating," whereby an oxygen atom is clamped between two voluminous metal atoms, (this effect is similar to that observed upon the "chelating" of bulky acceptors by metal atoms in biferrocenes and ferrocenoylruthenocenes [127, 128]).

| XXXVIII | XXXIX |

Thus, the results of References 105-111, 60 can be treated as confirming the proposed possibility of modifying the ferrocene geometry and properties by means of chelate formation. In this case, the systems containing two chelatophoric substituents in both cyclic ligands are similar, (upon their closing via the coordination node), to ferrocenophanes and play the role of macrocycles. However, these complexes are a more convenient object for the investigation of conformational behavior of ferrocene, because a gradual variation of the tilt angle between the rings involves no difficulties in the organic synthesis.

Moreover, these studies open the possibility of an intentional synthesis of ferrocene-containing macrocyclic heterometallic complexes where the character and size of macrocyles are controlled by the radius of a complex-forming metal, composition, and linking of the metallochelate cycle.

(Translated by P.P. Pozdeev)

REFERENCES

1. T.J. Kealy and P.L. Pauson, "Nature," 168 (1951), 1039.
2. S.A. Miller, J.A. Tebboth and J.F. Tremaine, *J. Chem. Soc.*, No. 2 (1952), 632.
3. A.N. Nesmeyanov, "Selected Works. Ferrocene and Related Compounds," Moscow: Nauka, 1982) (in Russian).
4. A.N. Nesmeyanov, Selected Works "Ferrocene Chemistry," (Moscow: Nauka, 1969) (in Russian).
5. M.D. Reshetova, "Ferrocene Applications in Industry," (Moscow: Izd. NII TEKhIM, 1975) (in Russian).
6. M.D. Reshetova, "Applications of Ferrocene," (Moscow: Izd. NII TEKhIM, 1976).
7. N.S. Kochetkova, and Yu.K. Krynkina, *Uspekhi Khim.*, 47, No. 5 (1978) 934.

8. M. Green, "Organometallic Compounds," Vol. 2, The Transition Elements (London: Methuen, 1968).
9. H.M. Colhown, D. Holton, D. Thompson, and M. Twigg., *New Pathways in Organic Synthesis. Practical Application of the Transition Metal Complexes)*
10. Ya.M. Paushkin, T.P. Vishnyakova and I.D. Vlasova, *Zh. Org. Khim.*, 39 (1969) 2379.
11. E.G. Kasting, H. Naarman, H. Reis and C. Berding, *Angew. Chem.*, 77, No. 7, (1965) 313.
12. E.I. Edwards, R. Epton and G. Marr, *J. Organometal. Chem.*, 85, No. 2 (1975) 23.
13. E.I. Edwards, R. Epton and G. Marr, *J. Organometal. Chem.*, 107, No. 3 (1976) 351.
14. A.G. Osborne and R.H. Whitley, *J. Organometal. Chem.*, 193, No. 3 (1980) 345.
15. A.G. Nagy, J. Dezzsi and M. Hillman, *J. Organometal. Chem.*, 117, No. 1 (1976) 55.
16. R. Dagai, *Chem. Eng. News.*,60, No. 9 (1982) 23, 24, 27.
17. C.R. Hauser and J.K. Lindsey, *J. Org. Chem.*, 22, No. 5 (1957) 482.
18. L. Wolf, M. Beer, *Naturwiss.*, 44, No. 16, (1957) 422.
19. V. Weinmar, *Naturwiss.*, 45, No. 13 (1958) 311.
20. C.R. Pedersen and V. Weinmar, Pat. 2875223 (U.S.A.).
21. F. Dietze, E. Butter and E. Uhlmann, *Z. Anorg. Allgem. Chem.*, 400, No. 4 (1973) 51.
22. H. Henning, O. Gurtler, *J. Organometal. Chem.*, 11, No. 2 (1968) 307.
23. O.N. Suvorova, "Synthesis and Study of Properties of Iron Metalloorganic Derivatives Containing b-diketone Groups." Thesis. (Gorky, Univ. Press, 1970) (in Russian).
24. H. Henning, O. Gurtler, *J. Electroanal. Chem.*, 30, No. 2 (1971) 253.
25. J.-C. Lu, Y.-C. Zhang, J.-H. Bian, *Hua Shueh Tung Pao*, No. 2 (1981) 88; c.a., 95 (1981) 24007v.
26. H. Imai and Y. Yosikashi, Nippon Kagaku zhasshi., *J. Chem. Soc. Japan, Pure Chem. Sect.*, 91 No. 5 (1970) 452, A26.
27. H. Imai and T. Ota, Bull. Chem. Soc. Japan, 47, No. 10 (1974) 2497.
28. G. Zhang, H. Xiong and L. Zhang in: "New Frontiers Rare Earth Sci. and Appl. Proc. Int. Conf.," (Beijing, Sept. 10-14, 1985) Vol. 1, p. 191.
29. A.M. Shevchik, Yu.P., Losev and Ya.M. Paushkin, in: "Structure, Properties and Applications of Metal b-diketonates," ed. by V.I. Spitsin (Moscow: Nauka 1978), p. 136 (in Russian).
30. Yu.G. Orlik, P.N. Gaponik, A.I. Lesnikovich and A.I. Vrublevskii *Zh. Org. Khim.*, 48, No. 7 (1978) 1601.
31. H. Imai and T. Shiraiwa, Technol Repts Kansai Univ., No. 22 (1981), 107.
32. K.C. Joshi, V.N. Pata, *Coord. Chem. Revs.*, 22, No. 1-2 (1977) 37.
33. I.I. Kirillova, Yu.T. Struchkov, L.I. Martynenko and N.G. Dzyubenko in "Metal b-diketonates," (Moscow: Nauka, 1978), p. 124 (in Russian).
34. S. Kawaguchi, *Coord. Chem. Revs..*, 70, No. 1-2 (1986) 51.

35. "Theoretical and Applied Chemistry of Metal b-diketonates," ed. by V.I. Spitsin and L.I. Martynenko (Moscow: Nauka, 1985) (in Russian).

36. U. Casselato, M. Vidali, G. Bandoli, P.A. Vigato and D.A. Clemete, *Inorg. Nucl. Chem. Lett.*, 9 No. 4 (1973) 299.

37. P.A. Vigato, U. Casselato, D.A. Clemente, G. Bandoli, M. Vidali, *J. Inorg. Nucl. Chem.*, 35, No. 12 (1973) 4131.

38. "Problems of Chemistry and Applications of Metal b-Diketonates," ed. by V.I. Spitsin, (Moscow: Nauka, 1982) (in Russian).

39. G.J. Bullen, *Nature*, 177, No. 45007 (1956) 537.

40. G.J. Bullen, R. Mason and P. Pauling, *Nature*, 189, No. 4761 (1961) 291.

41. Pat. No. 1251954 (W. Germany).

42. G.B. Kazarinov, O.N. Vylegzhanina, G.A. Domrachev and G.A. Razuvaev, in: "Methods of Preparation and Analysis of Special-Purity Substances," (Moscow: Nauka, 1970), p. 123 (in Russian).

43. G.A. Razuvaev, G.A. Domrachev, O.N. Suvorova and L.G. Abakumova, *J. Organometal. Chem.*, 32, No. 1 (1971) 113.

44. O.N. Suvorova, V.V. Sharutin and G.A. Domrachev, in: "Structure, properties and applications of metals b-diketonates," (Moscow: Nauka, 1978), p. 132 (in Russian).

45. B. Hetnarski, *Bull. Acad. Polonaise Sci., Ser. Sci. Chim.*, 13, No. 9 (1965) 563.

46. B. Hetnarski, *Bull. Acad. Polonaise Sci., Ser. Sci. Chim.*, 13, No. 9 (1965) 523.

47. B. Hetnarski, *Bull. Acad. Polonaise Sci., Ser. Sci. Chim.*, 13, No. 9 (1965) 557.

48. B. Hetnarski, *Dokl. Akad. Nauk SSSR*, 156, No. 3 (1964) 604.

49. M. Rosenberg, R.W. Fish and C. Bennett, *J. Amer. Chem. Soc.*, 86, No. 22 (1964) 5166.

50. R.L. Collins and R. Pettit, *J. Inorg. Nucl. Chem.*, 29, No. 2 (1967) 503.

51. O.W. Webster, W. Mahler and R.E. Benson, *J. Amer. Chem. Soc.*, 84, No. 19 (1962) 3678.

52. L. Wolf and H. Henning, *Z. Anorg. Allgem. Chem.* 341, Nos. 1-2 (1965) 1.

53. D.N. Henrickson, J.S. Sohn and H.B. Gray, *Inorg. Chem.*, 10, No. 8 (1971), 1559.

54. G. Wilkinson, M. Rosenblum, M.C. Witing and R.B. Woodward, *J. Amer. Chem. Soc.*, 74, No. 8, (1952) 2125.

55. "Hard and Soft Acids and Bases," ed. R.G. Pearson (Stroutsbourg, Penn., 1973).

56. R.G. Pearson, *Uspekhi Khim.*, 40, No. 7 (1971) 1241.

57. A.D. Garnovskii, A.P. Sadimenko, O.A. Osipov, and G.V. Tsintsadze, "Hard-Soft Interactions in Coordination Chemistry," (Rostovon-Don: Izdat. RGU, 1986).

58. Yu.N. Kukushkin, *Koord. Khim.*, 4, No. 8 (1978) 1170.

59. R.J. Balahura and N.A. Lewis, *Coord. Chem. Revs.*, 20, No. 2 (1976) 109.

60. C.R. Hauser and C.E. Cain, *J. Org. Chem.*, 23, No. 8 (1958) 1142.

61. H.H. Wei and D.-M. Horn, *Transit. Met. Chem.*, 6, No. 6 (1981) 319.

62. P.N. Gaponik, A.I. Lesnikovich and Yu.G. Orlik, *Abstr. XII All-Union Chugaev Meeting on Chemistry of Complex Compounds* (Novosibirsk, 1975) Vol. 3, p. 502.
63. A.I. Vrublevskii, P.N. Gaponik, A.I. Lesnikovich and Yu.G. Orlik, *Abstr. XIII All-Union Meeting on Chemistry of Complex Compounds* (Moscow, 1978), p. 91.
64. J.P. Wilkins and E.L. Wittbecker, Pat. 2659711 (U.S.A.)
65. P.L. Pauson, *Quart. Revs. London Chem. Soc.,* 9, No. 4 (1955) 391.
66. A.J. Palevnik, *Inorg. Chem.,* 9, No. 11 (1970), 2424.
67. B.W. Rocket and G. Marr, *J. Organomet. Chem.,* 167, No. 1 (1979) 53.
68. G. Marr and B.W. Rocket, *J. Organomet. Chem.,* 147, No. 3 (1978) 273.
69. V.A. Kogan, V.V. Zelentsov, N.V. Garbalau and V.V. Lukov, *Zh. Neorgan. Khim.,* 31, No. 11 (1986) 2831.
70. R.L. Dutta and M.M. Hossain, *J. Sci. Ind. Res.,* 44, No. 12, (1985) 635.
71. S. Padhyl and G.B. Kauffman, *Coord. Chem. Revs.,* 63, No. 1 (1985) 127.
72. D.M. Wiles and T. Suprunchuk, *Can. J. Chem.,* 46, No. 11 91968) 1865.
74. W.D. Fleischmann "Spektroscopische und Polarographysche Untersuchungen an Metallokomplexen des Formylferrocen Thiosemicarbazones." Diss. Dokt. Naturwiss., Fak. Allg. Wiss., Techn. Univ., Munchen, 1972.
75. W.D. Fleischmann and H.P. Fritz, *Z. Naturforsch.,* 28B (1973) 383.
76. Y. Omote, R. Kobayashi and N. Sugiyama, *Bull. Chem. Soc., Japan,* 46, No. 9 (1973) 2896.
77. A.P. Budhkar, U.N. Kantak and D.N. Sen. *Indian J. Chem.,* A16, No. 7 (1978) 626.
78. S.R. Patyl, U.N. Kantak and D.N. Sen., *Inorg. Chim. Acta.* 63, No. 2 (1982) 261.
79. S.R. Patyl, U.N. Katnak and D.N. Sen., *Inorg. Chim. Acta.,* 68, No. 1 (1983) 1.
80. P. Gouzerh, Y. Jeannin and V. Marechal, *Transit. Met. Chem.,* 9, No. 12 (1984) 482.
81. G.M. Larin, Z.M. Musaev, O.F. Khodzheav, S.D. Nasirdinov, V.V. Minin, V.G. Yusupov and N.A. Paripev, *Coord. Khim.,* 8, No. 10 (1982) 1329.
82. Z.M. Musaev, O.F. Khodzheav, G.M. Larin and N.A. Parpiev, *Koord. Khim.,* 8, No. 10 (1982) 1360.
83. M.F. Iskander, E.-Sayed and M.A. Lasheen, *Inorg. Chim.,* Acta, 16 (1976) 147.
84. A.J. Pearson, *Metalloorganic Chemistry* (Chichester: Wiley, 1985) p. 313.
85. T. Izumi, K. Endo, O. Saito, I. Shimizu, M. Maemura and A. Kasahara. *Bull. Chem. Soc. Japan,* 51, No. 2 (1978) 663.
86. V.I. Sokolov, L.L. Troitskaya and O.A. Reutov, *J. Organomet. Chem.,* 133, No. 2 (1977) 28.
87. S.S. Crawford and H.D. Kaesz, *Inorg. Chem.* 16, No. 12 (1977) 3193.
88. M. Nonoyama, *Inorg. Nucl. Chem. Lett.,* 12, No. 9 (1976) 709.
89. G.C. Gaunt and B.L. Shaw, *J. Organimet. Chem.,* 102, No. 4 91975) 511.

90. M. Sugimoto and M. Nonoyama, *Inorg. Nucl. Chem. Lett.* 15, Nos. 11-12 (1979) 405.
91. O.A. Osipov and V.I. Gaivoronskii, *Zh. Organ. Khim.*, 33, No. 4 (1963) 1346.
92. D.P. Graddon and G.M. Mockler, *Aust. J. Chem.*, 21, No. 6 (1968) 1487.
93. Ng.Ph. Buu-Hoi, Ng.H. Xuong, V.H. Vam, F. Binon and R. Royer, *J. Chem. Soc.*, No. 5 (1953) 1358.
94. E.J. Elsewsky and D.F. Martin, *J. Organomet. Chem.*, 5, No. 2 (1966) 203.
95. I.A. Eremina, in: "Oil, Gas and Their Products," (Moscow: Izdat. MINGKhP, 1971), p. 144 (in Russian).
96. I.A. Eremina, "Synthesis and Study of Polymer Chelate Compounds on the Basis of Ferrocene b-ketoimines," Thesis (Moscow. Izd. MINGKhP, 1973) (in Russian).
97. I.A. Eremina and A.A. Zavelova, in: "Oil, Gas and Their Products," (Moscow: Izdat. MINGKhP, 1971) p. 143 (in Russian).
98. T. Nomura, *Bull. Chem. Soc. Japan,* 56, No. 10 (1983) 2937.
99. R.D. ereman and D.P. Nalewajek, *J. Inorg. Nucl. Chem.*, 43, No. 3 (1981) 523.
100. B. Czech, A. Ratajczak and W. Szulbinski, *Pol. J. Chem.*, 55, No. 12 (1981) (1983) 2601.
101. M. Ertas, V. Ahsen, A. Gürek and O. Bekaroglu, *J. Organomet. Chem.*, 336, Nos. 1-2 (1987) 183.
102. B.V. Polyakov, I.G. Aganova and V.P. Tverdokhlebova in: "Abstr. XVI All-Union Meeting on Chemistry of Complex Compounds," (Krasnoyarsk, 1987) Vol., 1, p. 341 (in Russian).
103. M. Ertas, A.R. Koray, V. Ahsen and O. Bekaroglu, *J. Organomet. Chem.*, 317, No. 3 (1986) 301.
104. Z. Gang, L. Feng, Z. Jishan and M. Yongxiang, *Polyhedron,* 7 No. 5 (1988) 393.
105. A.N. Morozov, "Structural Features and Physico-Chemical Properties of Metallocene Based Metallocheelates," Thesis. (Rostovon-Don, 1988).
106. V.A. Kogan, V.V. Lukov and A.N. Morozov, in: "Abstr. IV All-Union Meeting on Spectroscopy of Coordination Compounds," (Krasnodar, 1986) p. 198.
107. V.A. Kogan, A.N. Morozov and V.V. Lukov, *Zh. Neorgan. Khim.*, 33, No. 1 (1988) 133.
108. V.V. Lukov, A.N. Morozov and V.A. Kogan, in: "Abstr. IX All-Union Meeting on Mathematical and Physical Methods in Coordination Chemistry," (Novosibirsk, 1987) Vol. 2, p. 101 (in Russian).
109. V.A. Kogan, A.N. Morozov and V.V. Lukov, *Zh. Neorgan. Khim.*, 33, No. 3 (1988) 655.
110. V.A. Kogan, A.N. Morozov, V.V. Lukov, A.Z. Rubezhov and A.A. Bezrukova in: "Abstr. IV All-Union Meeting on Chemistry of Non-Aquous Solutions of Inorganic and Complex Compounds," Rostov-on-Don, 1987 (Moscow: Nauka, 1987), p. 229 (in Russian).
111. V.A. Kogan, V.V. Lukov, A.N. Morozov, A.Z. Rubezhov and A.A. Bezrukova, in: "Abstr. IV All-Union Meeting on Chemistry of Non-

Aqueous Solutions of Inorganic and Complex Compounds, Rostov-on-Don," 1987 (Moscow, Nauka, 1987), p. 274 (in Russian).

112. N.N. Kharabaev and V.A. Kogan, *Dokl. Akad. Nauk SSSR*, 282, No. 2 (1985) 396.
113. V.A. Kogan, N.N. Kharabaev, O.A. Osipov and S.G. Kochin, *Zh. Strukt. Khim.*, 22, No. 1 (1981) 126.
114. I.C. Paul, *Chem. Communs.*, No. 12, (1966) 377.
115. Yu.T. Struchkov, *Zh. Organ. Khim.*, 27, No. 8 (1957) 2039.
116. K. Shlegel, *Uspekhi Khim.*, 49, No. 8 (1970) 1424.
117. W.H. Morrison Jr. and D.N. Hendrickson, *Inorg. Chem.*, 11, No. 12 (1972) 2912.
118. J.W. Lauher and R. Hoffmann, *J. Amer. Chem. Soc.*, 98, No. 7 (1976) 1729.
119. D.I. Kochubei, *Uspekhi Khim.*, 55, No. 3 (1986) 418.
120. J. Goulon and C. Goulon-Ginet, *Pure Appl. Chem.*, No. 12 (1982) 2307.
121. J.D. Dunitz and L.E. Orgel, *Nature*, 171 (1953) 121.
122. R.D. Shannon, *Acta Cryst.*, A32, No. 5 (1976) 551.
123. M.A. Ali, S.M.G. Hossain, S.M.M.H., Majumder and M.N. Uddin, *Polyhedron*, 6, No. 7 (1987) 1653.
124. B.A. Gingras, T. Suprunchuk, O. Bernardini, and C.-H. Bayley, *Can. J. Chem.*, 41, No. 6 91963) 1629.
125. A.A. Pendin, P.K. Leont'evskaya, T.L. Suverneva, I.V. Rozenkova and B.P. Nikolskii, *Dokl. Akad. Nauk SSSR*, 293, No. 6 (1987) 1411.
126. R. Singh and J. Subramanyan, *J. Sci. Ind. Res.*, 40, No. 1 (1981) 24.
127. M. Watanabe, I. Motoyama and H. Sano, *Bull. Chem. Soc.* Japan, 59, No. 7 (1986) 2103.
128. M. Watanabe, I. Motoyama and H. Sano., *Bull. Chem. Soc. Japan*, 59, No. 7 (1986) 2109.

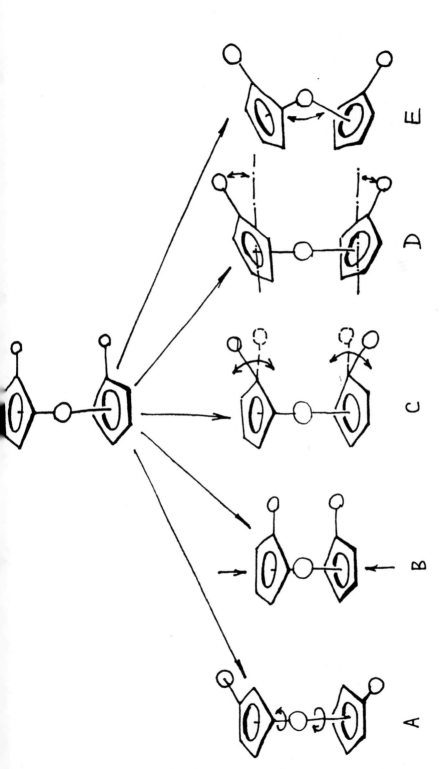

Figure 1. Possible types of sandwich deformations.

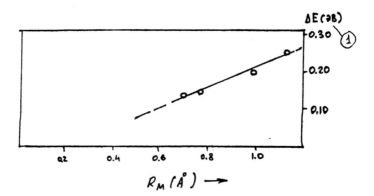

Figure 2. Variation of the d-d transition energy in ferrocene upon the chelate distortion vs. the radius of a complex-forming metal. Key: 1 - eV

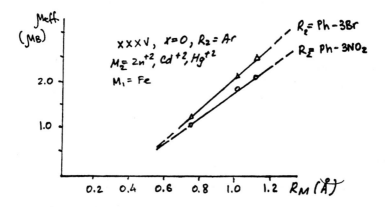

Figure 3. Effective magnetic moment of the XXXV type chelates vs. the radius of a complex-forming metal [122].

Cluster Compounds of Boron in Coordination Chemistry

N.T. *Kuznetsov and* K.A. *Solntsev*
Institute of General and Inorganic Chemistry

INTRODUCTION

Cluster compounds represent one of the most promising and rapidly developing sections of coordination chemistry.

In principle, the ability to form the cluster structure is inherent to the overwhelming majority of the elements possessing entirely or partially metal properties which include transition metals and so-called "semi-metals" such as boron, silicon, etc.

For about 40 years theoretical chemists and experimenters has been attracted to the vast class of polyhedral boron compounds whose skeleton either consists entirely of the atoms of boron or also includes the atoms of carbon, phosphorus, arsenic nitrogen and other transition and non-transition elements.

The specific features of these compounds likes in a high diversity and, sometimes, peculiarity of polyhedral structures and the ways of arrangement of atoms in polyhedron and their coordination. This is primarily caused by electron deficient cluster character of chemical bonds, high coordination numbers of the atoms composing the polyhedral skeleton and a high degree of delocalization of electrons in the skeleton.

The clearly-pronounced tendency towards diverse displacement and addition reactions and complex and unexpected transformations characteristic for these compounds opens wider opportunities for the synthesis of compounds and materials on the basis of their interest in theoretical and practical respects. It should be noted that the concepts developed for boron clusters and their derivatives, (carboranes, metal-boranes and metal-carboranes), turn out to be useful and applicable to a wide range of compounds, such as clusters of transition metals, p-complexes of metals and some hydrocarbon systems. It can be said that the chemistry of clusters is somewhere in-between the inorganic, (particularly coordination), element-organic and organic chemistry and, hence, embraces the problems inherent to these three sectors of chemical science. This can be illustrated by the example of the isoelectronic series $B_nH_n^{2-} - CB_{n-1}Hn^- - C_2B_{n-2}H_n$ the first of which is a purely inorganic system, the second-element-organic, and the third is distinguished by chemical C-H bond.

This paper naturally cannot embrace all the problems of the chemistry of cluster compounds of boron, and deals primarily with the chemistry of inorganic cluster structures and monocarbon isoelectronic analogs $CB_{n-1}H_n^-$ (n=6-12) and with the problems relating mainly to the use of the cluster compounds of boron as ligands or outer-sphere anions and coordination compounds. This is accounted for by the fact that the material included in the survey generalizes the results of the work conducted by the authors in this field for the past ten years.

Regarding the chemistry of carboranes $C_2B_{n-2}H_n$, metal-boranes and metalcarboranes, there are many rather detailed surveys in the literature and the author's contribution to these sections is considerably small [1-5].

1. GENERAL CHARACTERISTICS OF CLUSTER BORON COMPOUNDS

There are seven known boron-hydride structures which are the key structures for all other polyhedral compounds and are essentially two-valence systems $B_nH_n^{2-}$ (n=6-12) (Table 1).

Table 1.

Nos.	Ions	Polyhedron	Point group	Number of vertices	Number of edges	Number of faces
1	$B_6H_6^{2-}$	octahedron	Oh	6	12	8
2	$B_7H_7^{2-}$	heptagonal bipyramid	D5h	7	15	10
3	$B_8H_8^{2-}$	dodecahedron	D2h	8	18	12
4	$B_9H_9^{2-}$		D3h	8	21	14
5	$B_{10}H_{10}^{2-}$	3-cap trigonal prism	D4h	10	24	16
6	$B_{11}H_{11}^{2-}$	2-cap square antiprism	C2v			
7	$B_{12}H_{12}^{2-}$	octahedron icosahedron	Jn	11	27	18
				12	30	20

In the general form, the structure of polyhedral boron hydride anions satisfies the well-known King's conditions [6-7] for convex coordination polyhedrons containing only triangular (deltahedrons) and rectangular faces:

1. The Euler's relationship $1+2=v+f$, where
   l is the number of edges of the polyhedron;
   v is the number of vertices;
   f is the number of faces.
2. The relationship between the edges and faces $21=3t+4q$,
   where t is the number of deltahedrons;
   q is the number of rectangular faces.
3. The relationship between the edges and vertices $21=\sum_i ij$

   where i is the order of the vertex;
   j is the number of vertices of the order i.
4. The total number of vertices $V=\sum_i ji$.
5. The total number of faces $f=t+q$.

In a similar way the derivatives of these structures are constructed, including both those substituted over the exobonds B-H and those substituted over the skeleton. Sometimes, in place of one or a few atoms of boron, the skeleton includes atoms of other elements. In these cases a certain distortion of the polyhedron is observed, such as, when proceeding from $B_nH_n^{2-}$ to the isoelectronic carborane series $CB_{n-1}H_n^-$ and $C_2B_{n-2}Hn$ with the distortion being the greater the greater is the amount of the other heteroatoms which are introduced into the boron skeleton.

1.1. FORMATION OF POLYHEDRAL BORON HYDRIDE STRUCTURES

In the series $B_nH_n^{2-}$ not all the anions are equally accessible and studied. Whereas the anions $B_{10}H_{10}^{2-}$ and $B_{12}H_{12}^{2-}$ and pose no problems from the viewpoint of synthesis, storage and study, the compounds with the anions $B_{11}H_{11}^{2-}$ and $B_9H_9^{2-}$ are more difficult to synthesize and investigate, and the chemistry of the $B_6H_6^{2-}$ anion has been poorly studied until the publication of the authors' studies. Regarding the $B_7H_7^{2-}$ and $B_8H_8^{2-}$ anions, problems of synthesis are far from solved and their properties are not practically studied. On the basis of the theoretical generalization and the analysis of experimental data and basic factors governing the formation of one or another anion due to BH_x - condensation from nido-compounds or from other ready polyhedrons the authors have proposed the classification of the methods of synthesis of $B_nH_n^{2-}$ based on the separation according to the type of the original B - containing substances and the nature of the chemical bond in them [8]. The following three groups were proposed as the original substances for forming polyhedral structures:

- ionic salts (A), for instance MBH_4, MB_3H_8, $M_2B_{10}H_{10}$;
- neutral boranes (B): B_2H_6, B_5H_9, $B_{10}H_{14}$;
- adducts of boranes with Lewis bases (C)=for instance LBH_3, $B_{10}H_{12}L_2$ (Lewis base of type R_3N, etc.).

Under certain external conditions, (solvent, temperature, pressure), and the presence in the reaction system of the cations paired to the anion, (or in the case of possibility of their formation from the original substances), any paired combination of the original compounds incorporated in various groups or any compound belonging to this group may result in the formation of the polyhedral boron-hydride anions. This is accounted for by great stability of the polyhedral boron hydrides possessing a closed cellular structure and their high symmetry as compared with noncellular boron hydrides.

Original boron-containing compounds	Examples
I. Salt+adduct	$BH_4^- + Me_2S \cdot BH_3$; $B_{10}H_{10}^{2-} + Et_3N \cdot BH_3$
II. Salt+borine	$BH_4^- + B_2H_6$; $BH_4^- + B_5H_9$; $BH_4^- + B_{10}H_{14}$
III. Salt+borane	$Et_3N \cdot BH_3 + B_2H_6$; $Et_3N \cdot BH_3 + B_{10}H_{14}$
IV. Salt	oxidation $B_9H_9^{2-}$, Pyrolysis MB_3H_8
V. Adduct	pyrolysis $B_{10}H_{12}(Et_3N)_2$

It is evident that the formation of the monovalent polyhedral anions from neutral boranes is possible only in the presence of the components, ensuring the presence of cations in the reaction products, which reduces these processes to the HV reactions.

The proposed classification system embraces all known synthesis of polyhedral anions and, in addition, helps to select the direction of the search for new reactions of their formation. The versatility of such an approach was confirmed by such new formation reactions of polyhedral anions as transformation of tetrahydroborates of alkali metals of triethylamineborane into the salts of polyhedral boron hydride anions.

When studying the questions of competing reactions of formation of low and high polyhedral boron-hydride anions in the course of the latter's synthesis it has been established that the formation of, or an increase in, the yield of low polyhedrons due to a decrease in the yield of high ones requires the satisfaction of the following conditions: temperatures higher than those of the synthesis of the highest members of the series with n=10-12 and rapid withdrawal of the desired products from the reaction zone. The thermal transformations of salts with $B_3H_8^-$ anions has been studied as an example, the boric skeleton of which already contains deltahedrons, and the transfer to $B_6H_6^-$ requires the minimum number of interactions of the boron-containing particles. Depending on the conditions the thermal transformations of $B_3H_8^-$ result in $B_6H_6^{2-}, B_9H_9^{2-}, B_{10}H_{10}^{2-}$ and $B_{12}H_{12}$. The analysis of the temperature re-lationships G_{form}° for $B_6H_6^{2-}$ and $B_{12}H_{12}^{2-}$ has shown that, the yields of $B_6H_6^{2-}$ and $B_{12}H_{12}^{2-}$ at BH_x - condensation from $B_3H_8^-$ anion follow certain laws caused by a more rapid rise with temperature of the share of the entropic factor in the value ΔG_{form}° for forming the anion with a smaller number of atoms in the skeleton and by a simpler configuration which makes the prime formation of the low polyhedron possible. But it is known that the low polyhedrons due to BH_x - condensation are readily transformed into the high ones up to $B_nH_n^{2-}$. However, proceeding from $B_6H_6^{2-}$ to higher polyhedral ions these difficulties grow rapidly. The BH_9^{2-} anion has already been obtained by the solid-phase synthesis from $B_3H_8^-$ with the preset temperature regime being strictly observed.

1.2. STRUCTURE AND PECULIAR FEATURES OF CHEMICAL BONDS

It is generally assumed that the structures of most boron hydrides have the boric skeletons which, to some extent, can be regarded as fragments of icosahedron. This has become evident especially after defining the structures of the compounds with $B_nH_2^{2-}$ (n=6-12) and carboranes $CB_{n-2}H_n$ [9-12].

Williams has introduced the concept of three structural types, namely closo-, nido- and arachno-structures. In the close-structures the atoms of boron, (or with heteroatoms), are arranged in the vertices of closed polyhedrons with triangular faces (deltahedrons). These polyhedrons are the key for all other structural types. In the nido-structures, one vertex of the polyhedron remains unoccupied, and in arachno-structures two vertices are unoccupied.

The idea on multicenter two-electron bonds proposed by Longuet-Higgins first for the simple hydrides is very fruitful for these classes of compounds. Then, this idea was further developed for other classes of chemical compounds. This idea has been developed by Lipscomb into the topological method of localized bonds which has made it possible to not only explain, but also predict the new structures which subsequently have been experimentally realized [9]. However, this method wasn't that effective for the $B_nH_n^{2-}$

closo-structures since, for them, it is difficult to select only the set of localized bonds satisfying their symmetry on the basis of the equations of balance of electrons, charges and chemical bonds. The way out was essentially shown by Lipscomb himself, the use of the MO LLCAO method, i.e. instead of distribution, (localization), of the skeletal electron pairs it is necessary to compare their number with the number of the skeletal bonding MOs. For instance, in the $B_nH_n^{2-}$ closo-structures every BH fragment introduces three AOs and two electrons into formation of the skeleton, with two p-AOs being tangentially oriented with respect to the sphere circumscribed around the polyhedron, and one AO is directed towards the center of the polyhedron. In the formation of he polyhedron the exopolyhedral orbitals are not taken into account, so the interaction of 2n of the tangential AOs results in n-bonding and n-antibonding orbitals. Summing up all the bonding MOs of the skeleton, (with due account taken of the inner orbinal), gives their total number equal to n+1. The most stable closo-system exists when all the bonding MOs are occupied by electron pairs, which requires 2n+2 electrons. Since every BH fragment introduces one electron pair into the skeleton, the B_nH_n group cannot exist as a neutral molecule but can only exist as a two-valent anion. This postulate concerning the necessity to have 2n+2 electrons for bonding the closo-skeleton was termed the "2n+2" rule or the "Wade's rule" [16, 17]. It is worth noting the great contribution made to the solution of theoretical problems of the polyhedral structures by Mingos, Williams and Rudolph [18-23]. Since it is the symmetry of the polyhedron that determines the number of the bonding MOs, these closo-systems are the key for other isoelectron systems such $CB_{n-1}H_n^-, C_2H_{n-2}H_2$ and others. Thus, the CH group introduces three elements into the skeleton and, hence, with its introduction into the skeleton the valence decreases by unity as compared with $B_nH_n^{2-}$, and an introduction into the boric skeleton of two CH groups the neutral molecule $CB_{n-2}H_2$ is formed. In case of excess quantity of the skeletal electrons, since it is impossible for all of them to occupy the bonding MOs of the polyhedron, the polyhedron opens forming the open nido- and arachno-structures. And, conversely, the deficiency in the skeletal electrons results in the formation of very peculiar structures, the so-called "superelectron-deficient" ones where one or a few triangular faces are centered by additional vertices (caps). They can also be referred to as the closo-systems, since they contain only deltahedrons.

These concepts are also applicable to the mixed cluster structures whose polyhedrons, in addition to the atoms of boron (and carbon), contain the atoms of transition and non-transition metals-metal boranes and metal carboranes. The major condition here is retaining the polyhedral structure. The number of the electrons delivered into the skeleton depends on the character of the fragment being introduced into the skeleton, in which case the number of the electrons which the atom A or group AB delivers to the skeleton in the generalized form can be expressed as a+b-2, where a is the number of the electrons at the valence level of the atom A and b is the number of the electrons delivered by the atom B. On introduction into the polyhedron of the atom of the transition metal with five d-AOs which may contain ten electrons the afore-mentioned expression can assume the form of a+b-12, since its exopolyhedral capability shall be primarily utilized. However, when introducing heteroatoms into the boric skeleton the polyhedral structures may become distorted, with the severity of distortion being governed by the character of coordination of the heteroatom, (particularly, atom of metal), and by the orientation of its group with respect to the ligand.

Finally, the topological analysis makes it possible, for instance in such systems as $B_nH_n^{2-}$, to estimate the interatomic distances, (i.e. the lengths of the polyhedron edges, assuming that the skeletal atoms are bonded with one another by a set of individual two-center bonds directed along the polyhedron edges). The average bond order for a given closo-system can be calculated by dividing the total number of electron pairs (n+1) by the number of edges which for the n-vertex polyhedrons composed of deltahedrons is equal to 3n+6. This calculation gives the approximate order of the B-B bonds and their length in the skeleton of those tetrahedrons where all the edges are equal in length, (for instance for the octahedral anion $B_6H_6^{2-}$ the bond order is 0.58 and the interatomic distance is 1.69Å, in the icosahedron $B_{12}H_{12}$ these values are respectively equal to 0.43 and 1.77Å. For less symmetrical polyhedrons with n=8-10 the assumption that all the 2n+2 electrons are uniformly distributed among n atoms of the skeleton means that for every atom there shall be n+1/n electron pairs for the formation of the chemical bonds with x neighbors, (x is the skeletal coordination number), i.e. (n+1) electron pairs per each B-B bond. Thus, for each two-center bond between two adjacent atoms with coordination numbers x_1 and x_2 there shall be (n+1) $(x_1+x_2)/n\lozenge x_1 x_2$ electron pairs, which shall be anticipated bond order. For instance, this approach gives the following values of bond orders and interatomic distances = 0.50 (1.71Å) and 0.44 (1.89Å) in $B_9H_9^{2-}$ and 0.50 (1.73Å) ad 0.44 (1.83Å) in $B_{10}H_{10}^{2-}$.

In the first approximation, this method allows for the obtainment of rather correct values. Thus, the strongest bonds, (the shortest ones), are those linking the atoms of boron with the smallest coordination numbers, i.e. they carry a more negative change (as compared with all other atoms of boron) and the anion with the largest coordination numbers - more positive charge. This conclusion is generally in agreement with the quantum-chemistry analysis, though the length of the bonds in polyhedrons is determined not only by coordination numbers but by other factors as well.

A generalized topological rule, proposed by the Chinese scientists, is very useful and is applicable to both simple and conjugated polyhedral boranes and heteroboranes and accounts for many experimental facts [24, 25].

The number of the valence bonding orbitals VBO=4n-F(I), where n is the number of atoms of boron, and F is determined from the F=f+3(s+1)(II) equation, where f is the number of the deltahedrons of the closo-borane, s is the number of the atoms of boron removed (c-) from the corresponding closo-system or attached to it (c-). For instance, for the nido-systems S=-1, for the arachnosystems S=-2, for $B_6H_7^- = +1$ and so on. Thus, the VBO number is in full agreement with the real formulae and is determined only by the skeletal polyhedron. In the physical sense equation (1) means that in the formation of the boric skeleton F antibonding orbitals are formed, since F should be subtracted from the total number of the valence orbitals 4n which are formed from n atoms of boron.

Thus for the closo systems

Closo-system	F	4n-F	VF
$B_6H_6^{2-}$	11	13	26
$B_7H_7^{2-}$	13	15	30
$B_8H_8^{2-}$	15	17	34
$B_9H_9^{2-}$	17	19	38
$B_{10}H_{10}^{2-}$	19	21	42
$B_{11}H_{11}^{2-}$	21	23	46
$B_{12}H_{12}^{2-}$	23	25	50

This approach can also be applied to heteroboranes.

Since the polyhedral heteroboranes can contain the atoms of transition metals, each of which contains 9 valence orbitals, the equation becomes modified into:

$$VBO = 4n_1 + 9n_2 - \sum_{i=1}^{m} F_i + \sum_{i<j} U_{ij} - \sum_{i<j} V_{ij},$$

where n_1 is the number of atoms of the main subgroups (B, C, etc.);

n_2 is the number of atoms of transition metals;

U_{ij} is the number of the participating atoms of boron;

V_{ij} is the number of bonds between the i-th and j-th polyhedrons.

The quantum-chemistry calculations of E between the lowest unoccupied MO and the highest occupied MO for the closo-systems $B_nH_n^{2-}$ (n=6-12) show their relative stability [25].

Analyzing this relationship, Lipscomb et al. investigated the question of the boundaries of stability of polyhedral structures $B_nH_n^{2-}$ with n=12-24 and on the basis of the data for n=6-12 has come to the conclusion that stability depends, as a whole, on the number n, the charge of the anion and the coordination number of the atom of boron. From this point of view, the most stable are those with n=10 and 12, and n=14, 17.

The energy analysis confirms these conclusions: for n=22 the bond energy of one group BH decreases so rapidly that it makes the very existence of such polyhedrons problematic.

1.3. REACTION CAPACITY

The following type of reactions are characteristic for the polyhedral boron hydride anions:

- the reaction of the substitution of hydrogen for other atoms, groups or the whole molecules;
- the reactions of attachment of neutral molecules;
- the protonation reactions.

These reactions occur with the retaining of the boric skeleton. There are also a number of reactions where the boric skeleton is subjected to changes:

- the reactions of expansion and degradation of the boric skeleton;
- the reactions of insertion of heteroatoms into the boric skeleton.

The complex formation reactions shall be dealt with in separate sections.

It should be noted that the members of the $B_nH_n^{2-}$ series differ considerably in their thermal and chemical stability. Thus, the thermal stability of the boron clusters decreases in the series:

$$B_{12}H_{12}^{2-} > B_{10}H_{10}^{2-} > B_6H_6^{2-} > B_{11}H_{11}^{2-} > B_9H_9^{2-} > B_7H_7^{2-} > B_8H_8^{2-}.$$

Despite the electron-deficient character of the bond, it is sufficiently strong in absolute magnitudes and it reaches 600-800°C for $B_{12}H_{12}^{2-}$ (depending on the character of the cation). This is a unique property of the compounds of such a composition.

The kinetic stability of the compounds determined by the reaction rate varies, following approximately the same pattern decreasing in the series:

$$B_{12}H_{12}^{2-} > B_{10}H_{10}^{2-} > B_6H_6^{2-} > B_9H_9^{2-} > B_{11}H_{11}^{2-} > B_7H_7^{2-} \sim B_8H_8^{2-}$$

It should be noted that this series runs contrary to the contentions of some scientists suggesting the existence of a direct relationship between the stability and reactivity on the one hand and the number of boron anions in the skeleton on the other.

One of the important factors in this respect is a high symmetry of the anion deviation which results in differences in the distribution of electron density over the skeleton and in the appearance of non-equivalent boron atoms which in turn leads to the increasing reaction capacity.

The $B_nH_n^{2-}$ anions (particularly $B_{12}H_{12}^{2-}, B_{10}H_{10}^{2-}, B_6H_6^{2-}$), are readily halogenated by hydrogen halides, halogens and interhalogen compounds up to $B_nX_n^{2-}$ practically without destruction of the boric skeleton [26-34].

The rate of halogenation decreases in the F-Cl-Br-I series at transition from n=6 to n=12 as the atoms of hydrogen are progressively substituted by the halogen.

It should be emphasized that these halogenation processes are also applicable to $B_6H_6^{2-}$ where not only $B_6H_5X^{2-}$, (in the form of tetrabutylammonia salts), with X=Cl, Br, I are isolated, but even more deeply substituted $B_6H_{6-n}X_n^{2-}$, with them being synthesized with chlorine up to n=5, with bromine up to n=4, and with iodine up to only n=2 [35-41]. The process of denteroexchange in them is also studied in detail.

The thermal and kinetic stability increases in the series

$$B_nH_{n-1}H^{2-} \rightarrow B_nH_n^{2-}$$

$$B_nH_{n-m}I_m^{2-} \rightarrow B_nH_{n-m}Br_m^{2-} \rightarrow B_nH_{n-m}Cl_m^{2-} \rightarrow B_nH_{n-m}F_m^2$$

The halogenation process follows the stepwise course and depending on the conditions all the members of the $B_nH_{n-m}X_m^{2-}$ series can be isolated. The final substitution products $B_nX_n^{2-}$, even so large anions as $B_{12}I_{12}^{2-}$, are characterized by a higher thermal and chemical stability as compared with the initial $B_nH_n^{2-}$.

An interesting specific feature of the halogen-substituted compounds is their tendency towards the thermal processes of the type

$$Cs_2B_{10}Cl_{10} \xrightarrow{450-480°C} Cs_4B_{20}Cl_{16} \xrightarrow{540-590°C} Cs_2B_{12}Cl_{12}$$

When the hydrogen atoms are substituted with halogen the isomers are formed whose number differs from that calculated theoretically. Thus, in the system $B_{12}H_{12-x}F_x^2$ for $B_{12}H_{10}F_2^{2-}$ out of the three possible isomers only two, namely - 1.7 and 1.2 and the 1.12-isomer is absent, for $B_{12}H_8F_4^{2-}$ out of the 10 theoretically possible isomers only one - 1, 2, 8, 10-isomer is realized. In other words, only isomers which ensure the most uniform and symmetrical distribution of electron density over the polyhedron vertices are realized. This is true for highly-symmetrical cluster $B_{12}H_{12}^{2-}$ and $B_6H_6^{2-}$ structures, whereas in the cases of the clusters with non-equivalent atoms of boron of the

$B_{10}H_{10}^{2-}$ type, the substitution proceeds at the most reactionary centers, i.e. in the locations with minimum coordination numbers.

The hydrogen atoms in $B_nH_n^{2-}$ are capable of being substituted with other radicals-atoms or neutral molecules of the NR_3 type without changing the boric skeleton and charge [42].

For instance, under the action of mercury trifluoroacetate the $B_{12}H_{12}$ anion is subjected to the mercuration reaction with the formation of the substituted compound $[B_{12}H_{12-x}(HgCF_3COO)_x]$, where X=1-4, 6, 9±2 [43, 44].

Substitution of hydrogen with the group $HgCF_3COO^-$ proceeds with the formation of bonds B-Hg.

The intensity of the mercuration process is governed by the character of the cation - at mercuration of close-dodecaborates of alkali metals and barium the degree of substitution does not exceed four, with $Cs_2B_{12}H_{12}$ - three, for dodecaborates, three, and for tetraalcylammonia the process proceeds with high degrees of substitution up to the completely substituted $[B_{12}(HgCF_3COO)_{12}]^{2-}$.

Increasing concentration of the trifluoracetic acid in unhydrous media contributes to the formation of products with a high degree of substitution.

All the mercurated boron compounds are poorly soluble in water and most organic solvents with the exception of dimethylformamide and dymethylsulfoxides. Only completely substituted salt $[(CC_4H_9)_4N)]$ $[B_{12}(HgCF_3COO)_{12}]$ is sufficiently soluble in $CHCl_3$, $(CH_3)_2CO$, CH_2Cl_2 and CCl_4.

1.4. "SUPERELECTRONDEFICIENT" STRUCTURES

An interesting feature of the closo-boron-hydride $B_nH_n^{2-}$ structures is their ability to be protonated with the formation of a new class of "superelectrondeficient" structures.

The possibility of attaching the proton to the polyhedral closo-systems have been discussed by theoreticians many times beginning with the work of Evans in 1978, who proposed for $B_6H_6^-$ three possible configurations, the most promising among them in his opinion being the distorted octahedron with the bridging arrangement of the seventh atom of hydrogen between two atoms of boron - edges of the skeleton [46]. Brint [47,48] as a result of the analysis of $B_6H_7^-$ has come to the conclusion on preferable configuration in the form of the "cap" - type octahedron. These data were also confirmed by Dekock and Jasperse [51]. The similar calculations were also conducted by Wade et al. The quantum-chemistry analysis conducted by the authors for $B_6H_6^{2-}, B_6H_7^-; B_6H_8; Li_2B_6H_6$ and LiB_6H_7 [52] has shown that the most preferable configuration for the anion $B_6H_7^-$ is the "cap" - type octahedron with atom H_7 above the center of the face. The structure with H_7 above the edge lies by about 10 kcal higher and corresponds to the value of the migration barrier. For the internuclear distance, the value of 1.41Å was obtained in approximation CC/4-31G and the value of 1.25Å in approximation CC/OCT-3G. This difference is not surprising since the distances are not rigid and within the limits of 1.25 to 1.41Å the total energy varies inconsiderably.

For instance, in the case of $B_6H_6^{2-}$ the distances B-B and B-H calculated by the authors coincide with their experimental data to within 0.02-0.03Å. The calculated data show that the most probable migration path of the seventh proton is from face to face via the B-B edges. For the salt LiB_6H_7 and hypothetical hydride B_6H_8 the best structure has turned out to be the 2-cap octahedron in which the cation Li^+ (for LiB_6H_7) and proton H_7 are arranged above the opposite edges, with the presence of lithium resulting in the considerable deformation of the octahedron and increasing migration barrier. The B-B distances oriented to H_7 increase by 0.1Å, whereas in the opposite deltahedron they decrease. From the population analysis data in the isolated $B_6H_6^{2-}$ the atoms of boron have small negative and the atoms of hydrogen - small positive charge. In the $B_6H_7^-$ anion part of electron density is transferred from $B_6H_6^{2-}$ to H_7^+.

These theoretical calculations of the possibility of existence of the $B_6H_7^-$ anion and its anticipated construction have been successfully finalized by the synthesis of the compounds containing this ion and by the studies of their structure and properties. The synthesis has been conducted by the authors following the procedure described in [53].

$$B_6H_6^{2-} + (R_4)X \underset{PH=5}{\overset{HCl}{\rightarrow}} R_4B_6H_7 + X^-$$

$$R_4 = (CH_4)_4N^+; (C_2H_5)_4N^+; (C_4H_9)_4N^+; (C_6H_5)_4P^+$$

$$X^- = Cl^-; Br^-; I^-, NO_3$$

The salts containing $B_6H_7^-$ and such groups of cations as $[NiL_3]^{2+}$ where L-2.2' - bipyridine and 1,10 - phenanthroline were also synthesized. From the salts with metal cation only CsB_6H_7 has been obtained in a pure form.

The structure of the $B_6H_7^-$ anion has been proven by the authors by decoding of the structures of the following three crystals $[Ni(BiPY)_3](B_6H_7)_2$; $[(C_6H_5)_4P]B_6H_7$ and $[Ni(Phen)_3](B_6H_7)_2 \lozenge (CH_3)_2CO$, which has allowed the authors to make the final conclusion on the geometry of the $B_6H_7^-$ anion whose specific feature is the presence of the "cap" - type proton formally linked by the 4-center bond to three atoms of boron. The boron-boron distance in the deltahedron under this seventh proton is increased considerably (1.82Å) as compared with the other bonds in $B_6H_7^-$ (by about 0.12Å) and as compared with the boron-boron bonds in the initial $B_6H_6^{2-}$ (by about 0.1Å).

The electron density on H_7 is somewhat higher than all other end atoms of hydrogen. The B-H bond is also increased slightly as compared with $B_6H_6^{2-}$. The seventh atom of hydrogen ceases to migrate and is fixed on the edge at 190°C. The structure of the anion is confirmed by the spectral data - NMR spectroscopy "B" and "H", X-ray photoelectron spectroscopy, IRS. The method of dynamic NMR-spectroscopy has made it possible to determine the lifetime of the migrating proton above one of the edges of the octahedral boric skeleton (t) and the time of the quadruple relaxation of the atoms of boron bonded ($T°$) ad unbonded (T^c) with the face located atom of hydrogen [55]. The difference between the experimental and calculated NMR spectra of "B-'H did not exceed 5-8%. The obtained variation of t with temperature has made it possible to determine the energy parameters of the intramolecular migration of the seventh proton from one edge of the octahedral skeleton to another $H_{198}° = 48KJ/m$

$$S_{198}° = 45(10)J/m \text{ and } G = 39(2)KJ/m$$

The value of the constant of the scalar bond "B-'H=13Hz which is considerably smaller than $J''_{B-'H}$=146 Hz.

It should be noted that the protonation of $B_6H_6^{2-}$ results in a decrease in the thermal stability of the compounds, though the latter also depends on the character of the cation. Subjected to the protonation is not only the anion $B_6H_6^{2-}$, but also its halogen-substituted compounds such, $XB_6H_5^2 + H^+ = XB_6H_6^- (X = Cl, Br, I)$ with the salts of the composition $[(C_4H_9)_4N]XB_6H_6$[35-41] being isolated.

The $B_7H_7^{2-}$ and $B_8H_8^{2-}$ anions are rather unstable, which makes the experimental study of the processes of their protonation difficult. The quantum-chemistry analysis of the protonated forms $B_7H_8^-$ and $B_8H_9^-$ conducted by the authors has yielded the following results [57]. In the normal state of the anion $B_7H_8^-$ the additional proton is preferably located over the face of the bypyramid, but the potential barriers in the way of the proton migration are rather low, and as the temperature increases the proton can possibly move through the bipyramid edges, it is as if the proton is smeared over the equatorial region of the B_7 skeleton. It is here that the principal difference lies between this anion and B_6H_7, where the additional proton is free to migrate around the entire octahedron. Non-empirical analysis shows that the $B_8H_9^-$ anion turns out to be more rigid than $B_7B_8^-$ and more so than $B_6B_7^-$. The most preferable structure has turned out to be that where the ninth proton is located over the edge and, hence, the possibility of its migration over the skeleton is limited. Transitions from state I to state II and the oscillation of the proton along the polyhedron edge B_4B_4 are possible (at temperatures which are not too high), and a rise in temperature should make the migration of the additional proton in the direction perpendicular to the edge B_4-B_4 (barrier of about 10 kcal) easier. The character of the structural deformation of the $B_7H_8^-$ and $B_8H_9^-$ anions under the effect of the additional proton is somewhat similar to that observed in the B_6H_7 anion. The B-B bonds of the deltahedron which faces the additional proton become elongated and, as the distance to this proton increases, the geometrical distortion of the skeleton decays and on the opposite side of the polyhedron there may be observed the shortening of the B-B distances. This increase in the B-B distances as compared with the isolated $B_7H_7^{2-}$ and $B_8H_8^{2-}$ and amounts to 0.10-0.015Å depending on the tridentate or bidentate coordination of the additional proton. The calculations have shown that the additional proton draws part of the electron density away from the $B_7H_7^{2-}$ and $B_8H_8^-$ anions in the same manner as $B_6H_6^{2-}$.

The analysis of the B_9H_{10} anion has shown that the disturbing effect produced by the additional proton is such that the B_9 skeleton may radically change its structure by sequentially "opening" the sections adjacent to it and "closing" the opposite sections in synchronism with the migration over the skeleton, in other words, the mechanism of transformation of the closo-structure into the nido-structure facing at all times the additional proton. The difference in energies between the "starting" and optimized configurations was within 30 to 50 kcal/mol.

On the whole, the existence of the three following isomers of the $B_9H_{10}^-$ anion is possible, namely the basic anion, the isomer with the nido-structure with the energy lying higher by 9 kcal/mol and the configuration with the energy lying higher by 17 kcal/mol than the energy of the basic anion.

The authors' experimental data confirm these calculations, the isolation of $B_9H_{10}^-$ as difficultly-soluble compounds, in the same manner as it was done for $B_6H_7^-$ and $B_{10}H_{11}^-$ involves great experimental difficulties due to the instability of $B_9H_{10}^-$, where, in a number of cases the paramagnetic radical $B_9H_9^-$ is one of the decomposition products.

In the experiments conducted by the authors for obtaining the protonated form of the nonoborane anion in the presence of such large complex anions as $[Ni(Bipy)_3]^{2-}$ and $[Ni(Phen)_3]^{2-}$ [58] with the use of the "B and 'H NMR spectroscopy method the authors have managed to observe the formation of the $B_9H_{10}^-$ anion [ ] along with the decomposition products formed at all times.

The quantum-chemical analysis of the $B_{10}H_{11}^-$ anion predicts, as the basic configuration, the model where the eleventh hydrogen atom is located over the edge of the polar section of the polyhedron $B_{10}H_{10}^{-2}$, with the additional proton being nearer to the "cap" boron atom (interatomic distances 1.26 and

1.54Å). Compared to other protonated anions, this structure is distinguished by the fact that it exhibits the considerable extension of one base bond up to 2.04Å. In the basic structure there are three other configurations, II, III and IV, with the energies only differing slightly from one another (within 1 to 4 kcal). The proton in configuration II is located above the polar face of the polyhedron nearer to the "cap" boron atom, (1.30Å and 1.63Å), while in configurations III and IV two hydrogen atoms are located respectively above the "cap" boron atom and above the vertex of the antiprism.

Thus, the additional proton should be located in the vicinity of the "cap" boron atom and should migrate with very low barriers in the region over the edges and faces of the polar zone it is as if the proton is, "smeared" over the polar zone. The geometrical deformation of the anion $B_{10}H_{10}^{2-}$ at protonation is generally the same in character as that of the anion $B_6H_6^{2-}$. The B-B distance in the edges and faces of the polyhedron coordinated to the additional proton increases, with this increase being more pronounced than in $B_6H_7^-$ (where it is equal to 0.10-0.11Å when the proton is over the face and 0.14Å when it is over the edge). As the distance to the additional proton increases, the deformation of the B10 skeleton decreases. Just as for $B_6H_7^-$, in $B_{10}H_{11}^-$ the B-H distance remains practically unchanged as compared with the corresponding anions $B_6H_6^{2-}$ and $B_{10}H_{10}^{2-}$.

The use of the ideas developed for the anion about the stabilization of the protonated forms by means of large cations has led to the synthesis of the salts containing $B_{10}H_{11}^-$ [60]. For instance, the authors have obtained $[(C_6H_5)_4P]B_{10}H_{11}$ and $[(C_6H_5)_3PC_2H_5]B_{10}H_{11}$.

The spectral-analysis data confirms the presence of the $B_{10}H_{11}^-$ anion. For instance, the "B NMR-spectrum of the specimens at 183K contains two broad lines with d_1=25.67 and d_2=-23.50 m.p., with integral intensities being in the ratio of 1:9 and with the signal with 1 being split into the doublet with J_1=151.7 Hz. The authors suggest the following model with the geometry of $B_{10}H_{11}^-$ consistent with the data of NMR-spectrum. At temperatures close to 183K the additional proton is loosely bonded to one of the five coordination atoms of boron and, as it were, proceeds in the direction of the end B-H bond of the apical boron atom. This does not exclude its still weaker interaction with the four skeletal boron atoms nearest to the vertex. As the temperature increases, the process of exchange of the additional hydrogen atom between two identical apical boron atoms, (with coordination numbers 5), becomes initiated. This model suggests a considerable difference between the bond of the eleventh proton with the "apical" boron atom in the $B_{10}H_{11}^-$ anion and the conventional hydride end apical bond, which is confirmed by the "B NMR-spectra at 183K.

The authors have uncoded two crystalline structures, namely $[(C_6H_5)_4P]B_{10}H_{11}$(TFF $B_{10}H_{11}$) and $[(C_6H_5)_3PC_2H_5]B_{10}H_{11}$ (TFE $B_{10}H_{11}$).

The $B_{10}H_{11}^-$ anion in TFF $B_{10}H_{11}$ occupies its position on axis 4 and in the first approximation the form of the polyhedron is in agreement with the universally-adopted model for $B_{10}H_{10}$, however, the distribution of the interatomic distances gives no grounds, (distance B9-B9=2.11Å), for regarding the B-B bond as normal, (normal length is 1.75Å). Consequently, the upper part being formally a tetragonal pyramid contains at the base the atoms in coordination of 3 in terms of boron. Thus, there are the boron atoms of the three types, namely with coordination of 3 (four atoms), 4 (two atoms) and 5 (4 atoms) in terms of boron.

The other B-B distances in the $B_{10}H_{11}^-$ polyhedron lie within the range of 1.60 to 1.89Å. The cations and anions in the plane ab are packed in the staggered pattern formed along the axis in slightly corrugated layers.

The $B_{10}H_{11}^-$ anion in TFE $B_{10}H_{11}$ is of an unusual form that was not previously encountered among the closed boron-hydride polyhedrons, normally characterized by B-B-B trigonal faces. The characteristic feature of the new polyhedron is the presence of two adjacent rectangular faces the lengths of the diagonals of which 2.12; 2.31; 2.34 and 2.38Å suggests the absence of chemical bonds in these directions. This polyhedron with 14 faces and 22 edges can be described as two pentagonal pyramids with one common base side. The three other vertices of the base are arranged in the overlapped configuration and are bonded in pairs with the distances 1.70(2), 1.62(3), 1.43(3).

The lengths of the B-B bonds in the structure, except for the B_5-B_5 bond whose length is too short (1.43Å), range from 1.59 to 1.84Å. The coordination of the boron atoms in terms of boron is of two types: six vertices have the coordination 4, the remaining four - coordination 5. Thus, the peculiarity of the found polyhedron, which in the ideal case may have the symmetry mm 2, consists in the existence of two rectangular faces and much shorter distances, (13 distances are less than 1.75Å - the average B-B distance in other polyhedrons). The most prominent among these peculiar features if the extremely short B1-B5=1.42(3).Å bond whose length is regarded as unacceptable within the framework of prevailing ideas and available experimental data.

The cations and anions in the plane bc form the staggered arrangement, the corrugated layers are arranged along the axis a.

COMPARISON OF POLYHEDRONS $B_{10}H_{11}^-$ IN
TETRAGONAL AND MONOCLINIC STRUCTURES
AND TRANSITION BETWEEN THEM

The two forms of the $B_{10}H_{11}^-$ polyhedrons encountered in the tetragonal and monoclinic structures with the same number of vertices (10) differ (Table 5) in the number of bonds of the edges (20 and 22) and faces (12 and 14). From the viewpoint of crystallography these different polyhedrons can be transformed into each other with the use of simple operations. It is clear that the increase in the B-B distance up to 2.12 and 2.31Å which means the rupture of the B-B bond in the $B_{10}H_{11}^-$ as compared with the original $B_{10}H_{10}^{2-}$, occurs under the effect of the "additional" eleventh proton. Since only one proton is added, whereas the distance in $B_{10}H_{11tetr}^-$ increases in all the four B-B bonds of the base of one of the two pyramids, and in $B_{10}H_{11mon}^-$ - in one B-B bond of the base of one of the pyramids and in one B-B bond of the base of the second pyramid, it is regretful that in both cases the authors have failed to find the $B_{10}H_{11}^-$ anion completely ordered relative to the position of the additional eleventh proton. Of course, it is unlikely that the localization of such a proton would have been sufficiently reliable and therefore was not conducted. It should be noted that obtaining the compound with the ordered protonated hexaborate anion also involved considerable effort [53-55]. The only model common for both of the configurations is the model of the $B_{10}H_{11}$ anion with only one "large" distance B-B (2.11-2.31Å) or, which is one and the same, only with one rectangular face rather than two or four.

Within the framework of this idea about the structure of $B_{10}H_{11}^-$ the failure of the proton to be ordered relative to the fourth-order axis of the original cation $B_{10}H_{10}^{2-}$ leads to the structure of the $B_{10}H_{11tetr}^-$ anion and the failure of the proton to be ordered relative to the second-order axis of the original $B_{10}H_{10}^{2-}$ cation passing through the center of the edge connecting the bases of the both tetragonal pyramids leads to the structure of the $B_{10}H_{11}^-$ anion in TFE $B_{10}H_{11}$. If this is the case, then the very short B-B=1.43Å distance which reflects the infeasibility of accurate localization of two closely-spaced polyhedrons statistically juxtaposed on each other can be explained. Such analysis of the results of the X-ray structural analysis is in good agreement with the data of quantum-chemistry calculations [59] which predict as the basic configuration of $B_{10}H_{11}^-$ the configuration where the "additional" proton is located over the trigonal face of the tetragonal pyramid which results in the increase of the B-B distance of the base of this face up to 2.04Å. These X-ray structural analysis data are in good agreement with the model of the structure of the $B_{10}H_{11}^-$ anion proposed on the basis of the NMR-spectroscopy data. However, this is the second, although unclear, question as to whether this statistical disorder is static or dynamic.

In regards to the question as to which type of the polyhedrons-closo- or nido-systems - the $B_{10}H_{11}^-$ anion should be referred to, there is no unequivocal answer, though, most probably, this polyhedron is closer to the latter.

At last, the analysis of the $B_{12}H_{13}^-$ anion has shown that in terms of energy, it can assume one of the following three configurations: the tridentate and bidentate configurations where the additional proton is coordinated to the center of the face and center of the edge of the polyhedron, respectively and the configuration with two hydrogen atoms with one boron atom, with the latter configuration being most efficient, despite the fact that the difference in energy amounts to about 1 kcal. The latter configuration is distinguished by the unusual structure approaching more to the $B_{12}H_{11}H_2$ complex than to $B_{11}H_{11} BH_2$ with two B-H bonds at the one and the same atom of boron. This is confirmed by a large B-H distance equal to 1.47Å, and the H-H distance equal to 0.70Å is nearer to the distance in the isolated H_2 molecule. The bonding between the boron atom adjacent to the additional proton and the other boron atoms becomes somewhat stronger (by 0.04Å) as compared with the bonds in $B_{12}H_{12}^{2-}$, rather than weaker, as is the case in the entire series $B_nH_{n+1}^-$. The conclusion can be made that even if the $B_{12}H_{13}^-$ anion does exist, it ought to have a resistance to the decomposition with evolvement of the molecular hydrogen. In the course of studying protonation of $B_{12}H_{12}^{2-}$ the $B_{24}H_{23}^{3-}$ anion obtained earlier has been electrochemically synthesized. This anion is essentially two icosahedral polyhedrons bonded together be means of the B-H-B bond which is confirmed by the IR- and NMR-spectra.

Thus, on the basis of the theoretical studies and experiments the new property of the polyhedral $B_nH_n^{2-}$ anions (n=6-12) has been established, namely the tendency towards the protonation reactions with the formation of a new series of monovalent superelectrondeficient anions $B_nH_{n+1}^-$.

As a result of protonation the polyhedral anions are subjected to certain and specific changes. These new anions are distinguished by an unusual, unique character of bonding the additional proton with the boric skeleton and its migration. The theoretical analysis conducted by the authors has established the distinct ability to bond the "additional" proton is inherent to the $B_6H_6^{2-}, B_8H_8^{2-}, B_{10}H_{10}^{2-}$ anions and $B_{12}H_{12}^{2-}$, though the hydrogen migration path is different in all these anions.

In $B_6H_7^-$ the additional proton forms the 4-center bond with three boron atoms, thus forming the tetrahedral pyramid, located in the vertex. This proton migrates throughout the entire skeleton. In $B_8H_9^-$ the ninth proton preferably occupies the bridging position with respect to the pair of the neighboring 5-coordination boron atoms and its migration is essentially a pendular oscillations between the two deltahedrons adjacent to this edge. In $B_{10}H_{11}^-$, the additional proton is bonded with one of the apical boron atoms and two boron atoms of the base and, it is as if this proton is processing around the apical bond B-H.

Table 5. Comparative characteristics of polyhedrons

Number of faces	Number of bonds B-B edges	Number of vertices with co-ordination in terms of B			Bond length, Å		
		3	4	5	B-B	B-B	B-B average
14	22	–	6	4	1.43	1.84	1.71
12	20	4	2	4	1.60	1.89	1.71
16	24	–	2	8			
15	23		4	6			

In the closo-borates with n=7,9 and 11 the protonated forms are characterized by the bonding of the additional proton in a manner similar to that which takes place in one of the three aforementioned ones.

As far as $B_{12}H_{12}^{2-}$ is concerned, it can be regarded as saturated with respect to the protonation process.

The processes of protonation of the $B_nH_n^{2-}$ anions can be considered from the viewpoint of complex formation, since the formation of the bond between the skeleton and the additional proton is essentially the process of donor-acceptor interaction. Since this process takes place in the electron-deficient structures it is quite logical that the bond of the additional proton is distinguished by multicenter character. It is because of this unique analogy that the protonation process attracts so much attention.

2. COMPLEX COMPOUNDS OF TRANSITION METALS WITH OUTER-SPHERE POLYHEDRAL ANIONS

Aside from the fact that the polyhedral boron compounds a very interesting class of coordination compounds (just like the other types of cluster compounds), in many respects they considerably affect the coordination sphere of metals, acting as either outer-sphere anions or as specific ligands.

Used as the outer-sphere anions where the most stable representatives $B_{10}H_{10}^{2-}$ and $B_{12}H_{12}^{2-}$ and their halogen derivatives $B_{10}X_{10}^{2-}$ and $B_{12}X_{12}^{2-}$ (X=Cl, Br, I) and monocarboxylic isoelectron analogs with $B_nH_n^-$ (n=9,11).

The high kinetic and thermal stability of these anions in solution and in solid state and the unique possibility of varying the dimension and the charge of the ion, inaccessible for other anions, offers interesting opportunities for their use in synthetic coordination chemistry, particularly for investigating the effect of the geometrical and charge factors of the counterion of the coordination sphere of the transition metal.

2.1. COMPLEX COMPOUNDS OF PALLADIUM (II) AND PLATINUM (II)

Chloride complexes of Pd(II) and Pt(II) in the presence of $B_{12}H_{12}^{2-}$ and $B_{10}H_{10}^{2-}$ anions are decomposed with the isolation of free platinum and palladium. However, with halogen substituted anions $B_{12}X_{12}^{2-}$ and $B_{10}X_{10}^{2-}$ (X=Cl, Br) in the presence of L ligands (L - phenanthroline (Phen), bipyridil (Bipy) and triphenylphosphine (PPh3)... the following complex compounds are formed:

$$[PdL_2]\,BnXn \qquad [PtL_2]\,BnXn$$

$$[PdL_4Cl_2]\,BnXn \quad [Pt_2L_4Cl_2]\,BnXn$$

$$[PdL_3'Cl]\,BnXn \qquad [PtL_3'Cl]\,BnXn$$

[61-63]

The formation of the complexes of platinum and palladium of a similar type proceeds at different rates, the palladium complexes are precipitated from the solutions almost instantaneously, whereas their platinum analogs are formed for 15 to 16 hours at a temperature of about 50°C, which can be used for the rapid isolation of these metals.

The polyhedral anions play the part of the outer sphere of the complexes and are not directly bonded with the central atoms. Their presence results in a sharp rise in the thermal stability of the complexes, (making it more than twice as high as that for the well-known compounds [PdPhenCl₂] [PtBipyCl₂]), and a decrease in chemical activity, they are resistant to oxidizers, reducers and concentrated acids and alkalies. For instance, Pd(II) and Pt(II) are not oxidized to Pd(IV) and Pt(IV) by such oxidizers as hydrogen peroxide. They are difficult to solve in well known solvents.

The spectroscopic characteristics of these complexes are similar in type and point to the conventional planar coordination of the ligands to the central atom through the nitrogen atom and to the presence of almost unchanged outer-sphere anions $B_nX_n^{2-}$. Regarding the complex cation $[PdL_2]^2$, the use of the polyhedral anions is possibly the only method for its isolation in the stable state, since the earlier-popular method for its obtaining as perchlorate does not yield the kinetically and thermally stable product. The usual products in such reactions are the polymers [PdLCl₂] and [PtLCl₂].

The reactions of the acidocimplexes of palladium and platinum are rather specific with triphenlphosphine in the presence of halogenated closoborate anions.

In the case of palladium, two types of the cation complexes are isolated, depending on the conditions, namely $[Pd_2(PPh_3)Cl_3 B_nX_n(I)$ and, in an excess of triphenylphosphine, $[Pd(PPh_3)Cl] B_nX_n(II)$ [61].

Unlike the palladium compounds, the acidocomplexes of Pt (II) in these conditions yield only a certain quantity of $[Pt(PPh_3)Cl]_2B_nX_n$. The platinum analogs (II) can only be synthesized by the reaction of cis- $[Pt(PPh_3)_2Cl_2]$ with $Ag_2B_nX_n$. In other words, the transitions $[L_2MX_2]+LL\rlap{/}{E}[L_3MX^+]X^-$ typical for Pd(II) are not typical for Pt (II).

A similar-type complex cation structure unaffected by the character of the anion and is in agreement with the well-known data - square for $[ML_3Cl]^+$ and dinuclear complex with chlorion bridges. The high thermal and chemical stability of the inner coordination sphere is worth noting, thus, in the compounds of type I, triphenylphoshine is not oxidized inside the sphere with the use of the well-known methods.

The reactions of allyl complexes of palladium with the compounds containing the polyhedral anions are of interest $B_nX_n^{2-}$ (n=10, 12) (X=Cl, Br).

Thus, when using as the original compound:

$$[(\eta^3 - C_3H_5)PdCl_2] + Ag_2B_{10}Br_{10} \rightarrow [(\eta^3 - C_3H_5Pd(RCN)_2]_2 B_{10}Br_{10}$$

were synthesized in acetonitrile or benzonitrile.

In the case of acetonitrile solutions

$$[(\eta^3 - C_3H_5)Pd(CH_3CN)_2]_2 B_{10}Br_{10}C_6H_6 (A)$$

was isolated, and from the benzonitrile solutions $[(h^3-C_3H_5) Pd(C_6H_5CN)_2]B_{10}Br_{10}$ (B) was obtained, with the benzene molecule being easily removed in the former case by minor heating or vacuuming. Use of the polyhedral anions makes it possible to obtain the unknown coordination sphere which along with the allyl group includes the nitrile ligands. It should be emphasized that, in all probability, it is this ratio between the geometrical dimensions of the anion and cation that allows isolation of such components, since the other $B_nX_n^{2-}$ do not yield similar results, the reagents are isolated from the unchanged solutions.

The structure of the complex (A) is well confirmed by the 'H NMR spectra in deuteroacetone and is also in good agreement with the laser Raman spectra within the region of valent oscillations (Pd-C_3H_5) and deformative oscillations of the group   and with the data of the X-ray crystal structure analysis of $[(h^3-C_6H_5) Pd(CH_3CN)_2]_2B_{10}Br_{10}\Diamond C_6H_6$ [64].

The structure of the crystal is formed by the complex cations $[(h^3-C_3H_5Pd(CH_3(N)_2]^+$, anions $B_nBr_{10}^{2-}$ and molecules C_6H_6 in the 2:1:1 ratio located at the normal van der Waals distances.

The allyl radical occupies two coordination sites, thus supplementing the coordination sphere of palladium to the square. The geometry of the anion is practically ideal and the crystal field does not cause noticeable distortions. The benzene molecule plays a considerable part in the structure, since it occupies the space between the disproportionate planar cations and ellipsoidal anions in such a way that the plane of the benzene molecule is perpendicular to the planes of the complex cations, and stabilizes the entire structure.

Thus, use of the polyhedral anions $B_{12}Cl_{12}^{2-}$, $B_{12}Br_{12}^{2-}$, $B_{10}Cl_{10}^{2-}$ and $B_{10}Br_{10}^{2-}$ enables one to obtain a high yield of stable cation complexes of platinum and palladium. The complex compounds are stabilized in such a way that they are characterized by considerably higher thermal stability as compared with the cation complexes ClO_4^-, PF_6^- and BF_4^-.

2.2. COMPLEX COMPOUNDS OF
MN(II), FE(II), CO(II), NI(II), CU(II) AND LN(III)

Transition metals of groups VII, VIII and I of the periodic table are the most typical complex formers exhibiting a tendency towards the formation of complex compounds with a wide class of ligands. It should be noted that their coordination to the central atom may be accomplished through various atoms-oxygen, nitrogen, sulfur, phosphorus, etc. From this viewpoint, it is interesting to study the effect of the outer-sphere polyhedral $B_nX_n^{2-}$ anions on the coordination sphere of these metals.

The complex compounds of the compositions $[ML_2] B_nX_n$ (M=Mn(II), Fe(II), Cu(II); n=10, 12; X=H, Cl, Br, L=2.2-bipyridil, 1, 10-phenanthroline) and $[ML_3]B_nX_n[M=Ni)II]$, Fe(III), Co(II)] which are essentially fine-crystalline substances stable in air, low-soluble in water, but soluble in dimethylformamide and disulphoxide are formed relatively easy in aqueous-organic solutions. Thus, complexes are characterized by relatively high thermal stability and resistance to strong acids, alkalies and substances with strong oxidazing and reducing properties.

Such complex compounds of Ni(II) and Co(II) as - $[NiBipy_3] B_nX_n$, $[NiPhen_3] B_nX_n$, $[Co(Phen)_3]B_nH_n$, $Co[(Bipy)_3] B_nH_n$ are studied in more detail. These compounds are primarily distinguished by a high stabilization of the inner coordination sphere in these compounds. For instance, in $[NiPhen_3]$ $B_{12}X_{12}$ (X=Cl, Br) two molecules of phenanthroline are detached only at 480-490°C and the last molecule at 530-550°C. This stability is unique for the complexes with organic ligands. Within the temperature range of 550 to 650°C there exists the unhydrous compound $NiB_{12}X_{12}$ whose destruction begins at temperatures higher than 800°C. The bipyridil complexes are decomposed in approximately the same way, though they do not exhibit a well-pronounced stepwise character of detachment of ligands. It should be noted that the thermal stability of these compounds is a linear function of the dimensions of the outer-sphere polyhedral anions. For the complexes with $B_nH_n^{2-}$ it is smaller by 150-200°C. For instance, the thermal decomposition of

[MnPipy$_2$] B$_{10}$H$_{10}$, [FeBipy$_2$]B$_{10}$H$_{10}$ begins at 310-330°C of [Ni(Phen)$_2$]B$_n$H$_n$ and [Co(Phen)$_3$B$_n$H$_n$] [66] - at 280-300°C and of [CuBipy$_2$]B$_{10}$H$_{10}$H$_2$O - at 200°C. The greater reaction capacity of the anion inducing at elevated temperatures the processes of reduction of cations of metals, particularly Cu(II), probably play a certain part in the latter cases as well. It should be noted that in all the complexes the decomposition of the ligand in its turn serves as a catalyst to the process of decomposition of the polyhedral anion, particular B$_n$H$_n^{2-}$.

The ethylenediamine complex compounds of the [NiEn$_3$]B$_n$H$_n$ and [CoEn$_3$]B$_n$H$_n$(n=10, 12) type have turned out to be less stable.

The substitution of the B$_n$H$_n^{2-}$ anions in the complex compounds for their electron analogs CB$_{n-1}$ not changing their physical and chemical properties as a whole causes their thermal stability to rise considerably. For instance, the thermal stability [ML$_3$] (CB$_{n-1}$H$_n$)$_2$ (M=Co(II), Ni(II); L=Phen, Bipy; n=9,11) is higher by 80-100°C than that of the corresponding analogs [ML$_3$]B$_n$H$_n$, with the stability increasing with increasing n [65, 66] inside the series with CB$_{n-1}$X$_n^-$.

A considerable effect of the polyhedral anions on the composition of the inner sphere in the complexes of this type is observed in the case of the ethylenediamine complexes Co(II) and Ni (II): whereas in the complexes Co(II) the compounds of the composition [CoEn$_3$] B$_n$H$_n$◊2H$_2$O and [CoEn$_3$] CB$_{n-1}$◊H$_n$◊2H$_2$O are always isolated, in the same conditions the interaction of ethyleneamine with nickel salts in the presence of B$_n$H$_n^{2-}$ and CB$_{n-1}$H$_n^-$ takes a different course. In the cases with B$_n$H$_n^{2-}$ the complexes [NiEn$_3$] B$_n$H$_n$ are isolated, and in the cases with CB$_{n-1}$H$_n^-$ two types of complexes are formed concurrently, namely soluble complexes [NiEn$_3$] (CB$_{n-1}$H$_n$)$_2$◊2H$_2$O and insoluble complexes [NiEn$_2$(H$_2$O)$_2$](CB$_{n-1}$H$_n$)$_2$.

It should be noted that the compounds [NiEn$_3$]B$_n$H$_n$ are ionic. For instance, the analysis of the structure [NiEn$_3$]B$_{12}$H$_{12}$ has shown that each B$_{12}$H$_{12}^{2-}$ anion is coordinated with 5 cations [NiEn$_3$]$^{2+}$ in the form of a regular trigonal bipyramid. The cation [NiEn$_3$] is coordinated in the same way by the anion B$_{12}$H$_{12}^{2-}$. The average length of the Ni-N bonds is 2.138Å and of the B-B bonds is 1.786Å.

The similar properties are exhibited by the anions and CB$_{n-1}$H^-_n anions (n=10, 12) in the complexes of Co(II) and Ni(II) with benzoyl hydrazine C$_0$H$_5$CONHNH(BH) [67]. The interaction of the aquacomplexes [M(H$_2$0)$_6$]B$_n$H$_n$ and [M(H$_2$O)$_6$] (CB$_{n-1}$H$_n$)$_2$ with the ethanol solution of benzoyl hydrazine results in the precipitation of difficulty-soluble compounds of the composition [M(BH)$_3$]B$_n$H$_n$◊2H$_2$O and [M(BH)$_3$] (CB$_{n-1}$H$_n$)◊2H$_2$O: dihydrate - the compound stable in air, readily soluble in acetonnitrile, dimetrhylformamide and dimethylsulphoxide and difficulty soluble in water and nonpolar organic solvents [68].

The data of X-ray crystal structure analysis for $[Ni(BH)_3]B_{10}H_{10}\lozenge 2H_2O$ and $[Co(BH)_3] B_{12}H_{12}\lozenge 3C_2H_5OH$ and their spectroscopic and magnetic characteristics unequivocally confirm the octahedral coordination of the atoms of nickel (II) and cobalt (II) with the molecule of benzoylhydrazine being attached to the central atom bidentately due to the nitrogen atoms of the atmide group and the oxygen of the carbonyl group thus forming the five member ring. The molecules of benzoylhydrazine are in the amide tautomeric form, since the IR-spectra show a complete set of the frequencies of the valent and deformation oscillations of the coordinated NH_2-groups and there are no frequencies relating to the ordinary C-O bonds.

The complexes of Ni(II) and Co(II) containing $CB_{n-1}H_n^-$ form the coordination compounds containing their inner spheres along with the molecules of benzoylhydrazine two molecules of water - $[M(BH)_4(H_2O)_2](CB_{n-1}H_n)_2\lozenge H_2O]$. The specific feature of the coordinated benzoylhydrazine in these compounds is primarily its monodentateness. The coordinated benzoylhydrozine is attached to the central atom through the oxygen of the carbonyl group rather than through the M-N bond characteristic for the Co(II) and Ni(II) complex. In all probability, the sterical factors which require larger sizes of the complex cation and its definite coordination with the central anion play the key role in the formation of the inner coordination sphere. This is an example showing the effect of a decrease in the charge of the polyhedral ligand, with its geometry remaining unchanged, on the coordination sphere of the transition metal.

The same effect is also produced by an increase in the cation charge, for instance, in the lantanoid complexes of the type $[Ln(C_6H_5CONHNH_2)_4]_2$ $(B_{12}H_{12})_3$ where Ln=La, Nd, Sm, Dy, E, Vb, Lu [69].

Using the substituted benzoylhydrazine-meta-nitro-benzoylhydrazine as ligands leads to the formation of only one type of compounds with common formulae in all the afore-mentioned cases $M(m-NBH)_4(H_2O)_2B_nH_n\lozenge 2H_2O$ and $[M(m-NBH)_4(H_2O)_2](CB_{n-1}n)_2\lozenge 2H_2O$, where M-Co(II), Ni(II), m-NBH-m-$NO_2C_6H_4CONHNH_2$; n=10,12.

In structure and properties they are very close to the similar complexes with benzoylhydrazine. Despite the presence of a large number of the donor atoms both of oxygen and nitrogen, in all cases without exception the ligand is coordinated with the central atom through the oxygen of the carbonyl group.

The thermo-oxidative destruction of all the complexes of cobalt and nickel with Bh and m-NBH containing the outer-sphere anions $B_nH_n^{2-}$ and $CB_{n-1}H_n^-$ occurs within the temperature range of 200 to 350°C with subsequent destruction of the anions within the temperature range of 400 to 700°C, with the thermal stability increasing when proceeding from $B_nH_n^{2-}$ to $CB_{n-1}H_n$ and from n=10 to n=12. The specific feature of decomposition of the complexes with methanitrobenzoylhydrazine is their explosive character which is accounted for by the presence in one and the same molecule of the groups with oxidative (NO_2) and reducing ($B_nH_n^{2-}$ with $CB_{n-1}H_n^-$) properties.

The spectral characteristics and magnetic measurements of the complexes confirm the octahedral coordination of the ligands around the central atom for the complex cations ML_3^{2-}, planar coordination - for ML_2^{2+}, absence of any bond whatsoever between M and $B_nH_n^{2-}$ (no absorption at 2000-3000 cm^{-1} typical for bridging B-H...M - bonds), the ionic character of the chemical bond between the complex cations and ML_2^{2+} and polyhedral anions.

In the case of the compounds $[ML_3]B_{10}H_{10}$ and $[ML_2]B_{10}H_{10}$ the IR-spectrum is characterized by certain peculiar features, pointing to the presence of the influence of the ligands (Phen and Bipy) and $B_{10}H_{10}^{2-}$ on each other: for instance, there are no strips of the valent oscillations of the B-H apical at 2520-2540 cm^{-1} and the strips of absorption of the equatorial B-H bond are split into 2 to 3 components.

The phenanthroline complexes containing molecules of water in the outer sphere are characterized by the formation of vast hydrogen bonds between the ligand and the molecule of water close to the phenanthroline (according to the IR-spectra data - the region within 600 to 750 cm^{-1}). This is a specific case of the hydride-proton interaction between hydrogens of the polyhedral anions and those of the ligands [66].

The study of the processes of formation of the phosphonium complexes $[Ni(PPh_3)_2 X_2]$ (X=Cl, Br, I) with the polyhedral anions $B_{12}X_{12}^{2-}$ (X=Cl, Br) in acetonitrile, has established the formation of rather stable acetonitrile complexes $[Ni(CH_3CN)_6] B_{12}Cl_{12}$ and $[Ni(CH_3CN)_6]B_{12}Br_{12}\lozenge CH_3CN$ unsoluble in most organic solvents, but readily soluble in water with decomposition to $NiB_{12}Cl_{12}\lozenge 6H_2O$ and $NiB_{12}Br_{12}\lozenge 6H_2O$. The spectroscopic analysis of the acetonitrile complexes confirms their octahedral structure and the presence of the Ni-N bond (230, 266 and 280 cm^{-1}). The values of magnetic susceptibility at 20°C (3.29 mB for $[Ni(CH_3CN)_6]B_{12}Cl_{12}$ and 3.18 mB for $[Ni(CH_3CN)_6] B_{12}Br_{12}\lozenge CN_3CN$ are typical for the octahedral nickel complexes.

The stabilization of the first coordination sphere manifests itself most vividly in its thermal properties. It is known, for instance, that the acetonitrile complexes $[NI(CH_3CN)_4]X_2$ (X=Cl, Br, I) are eroded with a partial loss of acetonitrile, at as low as room temperature, whereas the $Ni(CH_3CH_6)B_{12}X_{12}$ complex begins to detach the coordinated acetonitrile at a temperature as high as 290°C for $B_{12}Cl_{12}$ and 350°C for $B_{12}Br_{12}$.

It should also be noted that use of the $B_{12}X_{12}^{2-}$ anion leads to an increase in the coordination number of up to 6 and to chemical stabilization of the first coordination sphere $[Ni(CH_3CN)_6]^{2+}$. Of some interest is the in-sphere reaction of the coordinated acetonitrile with ethanol which follows the course:

$$R-C\equiv N+R'-OH \rightarrow R'-O-C=NH \rightleftharpoons R'-C-NH$$
$$\qquad\qquad\qquad\qquad\ R\qquad\quad OR$$

A number of the complex compounds are also known where the $B_nH_n^{2-}$ anion has other substituents and even fairly large groups. For instance, from $B_{12}H_{12}^{2-}$ there was obtained $[(C_2H_5)_4N]\,B_{12}H_n\,NH_2COOCH_3$ capable of serving as ligands in coordination compounds such, for instance, as those of lantanoids

$$6H(B_{12}H_{11}NH_2COCH_3)+\ Ln_2O_3\ \underset{H_2O}{\overset{70°C}{\rightarrow}}\ 2Ln(B_{12}H_{11}NH_2COCH_3)_3\Diamond 5H_2O$$

Ln=La, Pr, Nd, Sm, Dy, En, Gd, Ho, Er, Tm, Vb, Lu [10].

The afore-mentioned data strongly suggest the existence of the bond between the anion and the lantanoid through the oxygen of the carboxyl group which, though weaker, is retained even in the case of formation of the complex

$$Ln(B_{12}H_{11}NH_2COCH_3)_3 \cdot 5H_2) + PyO \xrightarrow[C_2H_5OH]{room\ t°} [Ln(PyO)_6](B_{12}H_{11}NH_2$$

$(COCH_3)_3$ Ln = Nd, Sm, Dy, Eu, Yb, Lu

PyO − pyridil; N − oxid. [10].

2.3. COMPLEX COMPOUNDS OF URANYL UO_2^{2+} AND U(IV)

The development of the coordination chemistry of uran has been characterized lately by the advances in the synthesis of new complex compounds with neutral ligands. In overwhelming majority of the uranyl complexes, the coordination sphere along with neutral ligands contains acido-ligands which strongly affect the process of the sphere's formation. The complex compounds of uran with neutral ligand in the first coordination sphere acting as outer-sphere anions normally contain perchlorate-, nitrate- and iodide-ions, i.e. the anions poorly coordinated by uran.

Use of the polyhedral anions, which are also classified as poor-coordinated ligands, opens vast opportunities for synthesis of new uran complexes with neutral ligands.

By exchange reactions, the uran aquacomplexes $UO_2B_{12}H_{12}\Diamond 11H_2O$, $UO_2B_{10}H_{10}\Diamond 7H_2O$, $UO_2B_{12}B_{12}Cl_{12}\Diamond 11H_2O$ and $UO_2B_{10}Cl_{10}\Diamond 6H_2O$ [11] were synthesized. From the aqueous solutions of the afore-mentioned compounds $UO_2B_{12}H_{12}\Diamond 6CO(NH_2)_2$, $UO_2B_{10}H_{10}\Diamond nCO(NH_2)_2$ (n=6,7); $UO_2B_{10}H_{10}\Diamond 6CO\ (NH_2)_2\Diamond 2H_2O$, $UO_2B_{12}Cl_{12}\Diamond 5CO(NH_2)_2$ $UO_2B_{10}Cl_{10}\Diamond 8CO(NH_2)_2$ [72] were isolated in the presence of carbomide.

The number of the carbamide molecules contained in the complex depends on the initial ratios. Crystallization of the aquacomplexes from dimethylsulphide solutions has led to the formation of the complexes of the composition $UO_2B_{12}H_{12}\lozenge6SO(CH_3)_2$ and $UO_2B_{10}Cl_{10}\lozenge6SO(CH_3)$ [73].

The carbamide complexes of uranyl can also be obtained by solid-phase synthesis, for instance, from $UO2B12H12\lozenge11H2O$ and $CO(NH2)2$ with isolated individual compounds with 1:3, 1:5, 1:6, 1:7 and 1:8 ratio - $UO_2B_{12}H_{12}\lozenge3CO(NH_2)_2\lozenge3H_2O$, $UO_2B_{12}H_{12}\lozenge5CO(NH_3)_2$, $UO_2B_{12}H_{12}\lozenge6(NH_2)_2$, $UO_2B_{12}H_{12}\lozenge7CO(NH_2)_2$ and $UO_2B_{12}H_{12}\lozenge8CO(NH_2)_2$ [74].

The spectroscopic analysis of these complexes and X-ray crystal structure analysis of $UO_2\lozenge5CO(NH_2)_2B_{10}H_{10}\lozenge2CO(NH_2)_2$ and $[UO_2\lozenge5H_2O]B_{12}H_{12}\lozenge6H_2O$ [75] unequivocally points to the fact that, in all these cases, the coordination sphere of uranyl includes five molecules of ligands and that the complex cation is a pentagonal pyramid $[UO_2L_5]^{2+}$. In the cases of carbamide and dimethylsulphide complexes the coordination sphere is primarily substituted by the $CO(NH_2)_2$ and $SO(CH_3)$ molecules and only in the case of their deficiency the first coordination sphere includes the molecules of water. The excess H_2O $CO(NH_2)_2$ and $SO(CH_3)_2$ molecules are arranged on the second coordination sphere and are bonded with the complex cations and with one another by the vast hydrogen bonds.

In all the cases, the ligands are coordinated to the central atom through oxygen, just as in the case of the complex compounds of transition metals of groups VI-VIII and I. As compared with other anions the polyhedral anions considerably augment thermal stability of coordination compounds (by 100-150°C). This is observed most clearly in the carbamide complexes whose decomposition begins from 250-350°C.

The outer-sphere molecules of water, carbamide and dimethylsulphoxide are removed at considerably lower temperatures.

Stabilization of the coordination sphere is accompanied by considerable changes in the uranyl group and by marked deformation of the polyhedral anions. For instance, in the uranyl group the distance is within 1.81 to 1.84Å and OUO=164.8 degree angle, which means that these parameters differ considerably from the known values for the complexes of uranyl with other atoms.

As a result, thermal stability decreases considerably as compared with all other complexes with closo-borates.

Despite the retaining of the coordination number UO_2^{2+} in compounds of this type (5) insertion of the $B_{12}H_{12}^{2-}$ anions results in the entry of an unusually high number of neutral molecules. Another specific feature of complexes of this type, (particularly with carbamides), is the capability of reducing the uranyl group UO_2^{2+} to U(IV) with the formation of the corresponding complexes of four-valent uran. Use of zinc amalgam as a reducer made it possible to isolate the following complex compounds $U(B_{10}H_{10})_2\lozenge9CO(NH_2)_2\lozenge2H_2$ $U(B_{12}H_{12})_2\lozenge10CO(NH_2)_2\lozenge H_2)$; $U(B_{10}Cl_{10})_2\lozenge9CO(NH_2)_2\lozenge2H_2O$ and $U(B_{12}Cl_{12})_2\lozenge8CO(NH_2)_2$ [76].

It has been found that in all these compounds the coordination number of uran is equal to eight (octahedral coordination). In many respects their properties are similar to those of the corresponding uranyl compounds, particularly from the viewpoint of thermal stability and deformation of polyhedral anions.

The studies were also conducted for investigating the effect of the substituents in the ligand molecule on the properties of complex compounds with polyhedral anions. Used as sample compounds here were $[UO_2L_5]$ $B_{12}H_{12}$ (L=monomethylcarbamide dimethylcarbamide, dimethylacetamide, diethlacetamide), $UO_2[CO[N(CH_3)_2]_4$ $B_{12}H_{12} \Diamond CO[N(CH_2)_2]_2$ - tetramethyl urea (TM) and dimethylformamide $\{UO_2[COHN(CH_3)_2\}_5$.

The synthesis of these compounds is similar to that described in the foregoing text for other complexes of uranyl with neutral ligands. The composition and structure of the complex compounds is well confirmed by spectroscopic analyses. As expected, in all the cases the coordination is ensured through the oxygen of the carbonyl group and all the ligands are monodentate. Introduction of substituents into the carbamide is accompanied by its growth in size and considerably weakens the effect of the coordination cation on the polyhedral anion (thus decreasing its deformation). The data of the X-ray crystal structure analysis of $[UO_2TMU_4]$ $B_{12}H_{12}$ show that the polyhedral anion $B_{12}H_{12}^{2-}$ is essentially an ideal icosahedron (average B-B distance is 1.782Å and the difference in the lengths of B-B bonds in all cases does not exceed s (d)). The degree of distortion of the icosahedron in this complex is much lower than in the "classical" salt $K_2B_{12}H_{12}$ [77].

The major peculiarity of this structure is the unusual coordination number of UO_2^{2+} equal to four, all four molecules of tetramethylcarbamide are arranged in the equatorial plane of uranyl and are coordinated through oxygen. For uranyl, only two complexes are known with such a coordination number. The coordination number of uranyl, observed in this compound, is probably accounted for primarily by the inability of the polyhedral anions to be coordinated with the fragments of the uranyl type and by the fact that $B_{12}X_{12}^{2-}$ creates geometrical prerequisites for coordination around uranyl of only four sufficiently large-volume ligands of tetramethylcarbamide.

The monocarbon analogs $CB_{n-1}H_4^{2-}$ can also serve as the outersphere anions in the complex compounds of uranyl.

3. COMPLEX CLOSO-DODECABORATE-HALIDES

One of the peculiar features of the chemical behavior of alkali-metal closo-dodecaborates soluble in water is their tendency to form complex salts of the type $M_2B_{12}H_{12} \Diamond MX$, (where X=Cl, Br, I), BH_4^-, CN^-, NO_3^- and some others [26, 78, 79]. The formed complex salts may contain water, but the ratio $M_2B_{12}H_{12}/MX$ is always equal to unity. These complex compounds may also contain rather large-volume complex cations $[CO(NH_3)_6]_2B_{12}H_{12}(JO_3)_4$, $[CO(NH_3)_6]_2$ $H_{12}NO_3 \Diamond 2H_2O$, $[CO(NH_3)_5NO_3]B_{12}H_{12}(NO_3)_2$,

$[CO(NH_3)_6]_2B_{12}H_{12}\lozenge 2H_2O$, $[Cr(NH_3)_5\lozenge H_2O]_2B_{12}H_{12}(NO_3)_4$, $[Co(NH_3)_5 NO_3]_2B_{12}H_{12}(NO_3)_2$, $[Cr\ En_3]_2\ B_{12}H_{12}(Cr_2O_7)_2$, $[Cr\ En_3]B_{12}H_{12}NO_3$ and some others [80].

For dodecaborates the authors have synthesized alkali metal salts of the type $M_2B_{12}H_{12}\lozenge MX$, where M=K, Rb, Cs, X=$NO_3$, BH_4^-, $Tl_2B_{12}H_{12}$ $\lozenge TEX$, where X= BH_4^-, NO_3^- [81, 82, 87].

There are very wide opportunities for forming the compounds of the similar type. Normally, they are precipitated by mixing the aqueous solutions of the initial components or crystallized from aqueous solutions by gradual evaporation of water.

It should be noted that the tendency towards the formation of complex compounds is characteristic not only for dodecaborates, but is also observed in the $B_{10}H_{10}$ anion, though in a considerably lesser degree: the authors have synthesized $M_2B_{10}H_{10}\lozenge MX$, where M=K, Rb, Cs; X= BH_4^-, NO_3^- $Tl_2B_{10}H_{10}$ $\lozenge TlNO_3$.

Salts with even more mixed complications are known, such as $[Co(NH_3)_5H_2O]_2\ B_{10}H_{10}(NO_3)_4$, $[Co(NH_3)_5(BrO_3)]_2\ N_{10}H_{10}\lozenge(BrO_3)_2$, $[Co(NH_3)_6]B_{10}H_{10}(ClO_2)_4$, $[Co(NH_3)_5NO_3]_2\ [B_{10}H_{10}(ClO_3)_2]$ [83-85].

At the same time, unlike $B_{12}H_{12}^{2-}$ the complex decaborate halides are unknown and attempts to synthesize them ended in the crystallization of mechanical mixtures of initial components.

The present survey presents the results of the systematic studies of the entire series of closo-dodecaborate-halides of alkali metals conducted by the authors. Since both the cation and anion parts of these compounds has the geometrical shape close to the spherical one, this fact has made it possible to utilize the crystal-chemistry characteristics for establishing relationships between the structure and physical and chemical properties of these compounds.

3.1. PHYSICAL AND CHEMICAL STUDIES OF SYSTEMS AND SYNTHESIS OF COMPLEXES

Obtaining complex closo-dodecoborate-halide complexes poses no sophisticated problems - the authors should only mix the aqueous solutions of the initial components - alkali metal dodecaborates and the appropriate halides. In the case of cesium salts with X=Cl⁻, Br⁻, I⁻ precipitation of the double salts is observed in all the cases, irrespective of their concentration and the ratio between them. The corresponding rubidium salts are isolated when use of a concentrated solution is made and in the presence of small excess (stoichiometric) of rubidium halide. The double potassium salts are isolated by crystallization from saturated solutions of closo-dodecoborate and halides of potassium (in the presence of double excess of potassium halide) [82]. However, the double salts are not isolated in the cases of using alkali metal fluorides and sodium halides for the system $K_2B_{12}H_{12}$.

To more exactly determine the regions of existence and isolation of closo-dodecaborate-halides the authors have studied the solubility in the systems $M_2B_{12}H_{12}$ - MX - H_2O at 25°C [83-85]. The investigation of these systems was of particular interest from other viewpoints as well, primarily for clearing out whether the composition 1:1 is the only one some other ratios between dodecaborate and halide exist. For instance, in the $K_2B_{12}H_{12}$-KCl-H_2O system [84], (this system has been selected since it is an extreme one for which the complex potassium dodecaborate-chloride has failed to be obtained), the solution isotherm consists of two branches - the branch of crystallization of KCl and the branch of crystallization of $K_2B_{12}H_{12}$. The X-ray analysis has shown that in the given system within the entire interval of concentration the solid phases are double-phase $K_2B_{12}H_{12}$+KCl. In all other cases it has been clearly established that the formation of complex dodecaborate-halides the solution isotherm consists of three branches - the branch of crystallization of the double salt and the branches of crystallization of initial components. In all the cases the complex dodecaborate-halides have only one composition - 1:1. The composition of the compounds may be represented in different ways - as the conventional formulae of double salts $M_2B_{12}H_{12}\Diamond MX$ or as $M_3B_{12}H_{12}X$. The data of the X-ray crystal structure analysis have shown that the only correct way of representing the formulae of these compounds is the second one, since all the metal atoms are structurally identical.

3.2. Structure and Crystal-Chemistry Stability Criterion

To confirm the unique character of closo-dodecoborate-halides of alkali metals, obtain the data for their identification and clear up the question of the isostructural character of the entire series the authors have conducted the detailed X-ray crystal structure analysis of these compounds with identification of the crystal structure of $K_2B_{12}H_{12}\Diamond KBr$ [86].

It has been established that all salts are crystallized in rhombohedral syngony, $R\bar{3}m$ and are isostructural with the parameters presented in the table, Z=1, density varies within a wide range from 1.76 to 2.86 g/cm^3.

Parameters of elementary cells of $M_2B_{12}H_{12}\Diamond MX$

X	Cl		Br		J	
M	a,Å	a,deg.	a,Å	a,deg.	a,Å	a, deg
K	–	–	6.891	93 18	7.005	93 52
Rb	6.965	93 06	7.014	93 12	7.139	93 34
	7.179	92 44	7.232	92 48	7.352	93 18

The boron atoms are located in the vertices of the rhombohedral cell, the potassium atoms, in the center of cell edges and in the center the icosahedral anion $B_{12}H_{12}^{2-}$ is located. Four anions are arranged around the potassium in the form of a square and two halogens supplement the square to obtain the octahedron. In turn, the potassium atoms form around the bromine the poly-hedron in the form of nearly regular octahedron. The $B_{12}H_{12}^{2-}$ anion has the shape of a distorted icosahedron. From the character of the distortion of the boron skeleton it is seen that the major effect on the anion is produced by metal ions forming the almost regular cubooctahedron around it. The distance B-B varies within 1.75 to 1.84Å, the distance B-H within 1.09 to 1.19Å.

The crystal structure $K_2B_{12}H_{12}\lozenge KB_2$ bears a certain resemblance to the structure of perovskite CaO_3Ti. The lower symmetry of the all is accounted for by the fact that the perovskite structure is based on the package of the spherical ions, whereas in the present case the package includes the ions of not only spherical but also of icosahedral form.

The analysis of the coordination polyhedrons and their geometrical sta-bility limits has established the following three most important factors de-termining the crystal-chemistry criteria for the boundaries of existence of the complex salts of this type:

1. The stability of the polyhedron around the potassium ion consisting of four $B_{12}H_{12}^{2-}$ anions and two bromine atoms is, to a considerable degree, de-pendent on the size of the cation the radius of which can be found from the following relationship:

$$_{cat} = R_{an} \cdot \sqrt{2} - R_{an} = 0.44\ R_{an}$$

Since the $B_{12}H_{12}^{2-}$ radius is equal to 3.23Å, then the minimum radius of the cation for which such a coordination is still stable is equal to 1.34Å. Since the radius of the potassium ion is equal to namely this value and the radius of the potassium ion is equal to namely this value and the radius of the sodium ion is much smaller, the existence of sodium dodecaborate-halides is infeasible in the structure of these compounds.

2. The similar analysis of the geometrical stability limits the octahe-dral surrounding of the halide-ion by the ions of the metal has made it pos-sible to draw the conclusion that all the halide-ions, including the fluoride-ion can act as the halide-ion. For this purpose, the cations of larger sizes are needed.

3. The important characteristic determining the boundaries of existence of dodecoborate-halides of alkali metals is the sum of the ionic diameters of the metal and halogen. This sum cannot be less than the value characteriz-ing the limit distance to which two $B_{12}H_{12}^{2-}$ anions can approach each other. Since at contact of the dodecaborate anion with the cations its ionic radius is equal to 3.23Å, then the distance between such two atoms in the crystalline structure cannot be less than the sum of their radii - 6.46Å. Hence the param-eter of the elementary cell (a) for the dodecaborate-halides should exceed this value $d_{M^+}+d_{X}\geq6.46Å$.

Values of sums of diameters $d_{M^{++}} + d_{X^-}$ (Å)

$M^+ X^-$	F^-	Cl^-	Br^-	J^-
Na^+	4.62	5.58	5.88	6.36
K^+	5.32	6.28	6.58	7.06
Rb^+	5.64	6.60	6.90	7.38
Cs^+	5.94	6.92	7.22	7.70

These calculated boundaries are in a good agreement with the experimental data, the absence of the dodecaborate-fluorides of alkali metals, sodium dodecaborate-halides and $K_2B_{12}H_{12} \lozenge KCl$. To obtain the dodecaborate-fluorides the ionic radius of the cation should be not less than 2Å.

The crystal-chemistry criteria are applicable only to the complex compounds containing the spherical-shape ions, and their application to the double salts containing the NO_3^-, CN^-, BH_4^-, etc. ions requires introduction of the appropriate corrections.

3.3. PHYSICAL AND CHEMICAL PROPERTIES AND THEIR RELATIONSHIPS WITH STRUCTURE

The isostructural character of the dodecaborate halides presupposes the similarity of their spectroscopic and other physical and chemical properties. The IR-spectra are fairly close to one another, and unlike the IR-spectra of the original closo-dodecoborates, the absorption strips characterizing the valent oscillations of the B-H bond (2480 cm^{-1}) are clearly seen to be separated from the valent-deformative oscillations (1080 cm^{-1}) of the boron skeleton B_{12} in the $B_{12}H_{12}^{2-}$ anion.

This is accounted for by lowering the symmetry of the crystals from cubic for $M2B12H12$ to rhombohedral for $M_2B_{12}H_{12} \lozenge MX$. Under the effect of the lower-symmetry surroundings the $B_{12}H_{12}^{2-}$ anion is subjected to deformation. Judging by the degree of splitting of the absorption strips, the rubidium and cesium dodecaborate-chlorides are characterized by a minimum deformation which is in a good agreement with the structural data according to which the rhombohedral angle a is minimum in the case of dodecoborate-chlorides. The closer this angle is to the right, the closer the coordination polyhedron from the metal cation around $B_{12}H_{12}^{2-}$ is to the regular cuboctahedron, and, hence, the lesser is the distortion of the icosahedrical $B_{12}H_{12}^{2-}$.

The temperatures at the beginning of the thermooxidative destruction of dodecaborate-halides vary within a narrow range from 500 to 570°C, which is considerably lower than the thermal stability of the initial $M_2B_{12}H_{12}$.

The series here follow their own pattern. For instance, if the temperature of the 1-st exo-effect is considered as the parameter characterizing the stability of the given crystalline structure, then for the cesium dodecaborate-halides this temperature is maximum for bromides. It should be noted that bromides are distinguished by other special properties in addition to that described above. In this context, the structure of the crystal lattice of $K_2B_{12}H_{12}$ should once again be referred to. The experimental data point to the unusual thermal factors, very high for the heavy atoms and relatively low for the boron atoms. Explanation of this unusual value lies in the difference between the sum of the ionic diameters $d_{K^+} + d_{B_2^-}$ equal to 6.58Å, (on the one hand), and the distance between the $B_{12}H_{12}^{2-}$ anions equal to 6.87Å (on the other hand). It is this difference in 0.29Å that can account for the relatively high thermal oscillations of the boron and potassium atoms. It is as if the electrostatic interaction between K^+ and B_2^+, which tends to bring them closer together to the distance equal to the sum of the ionic radii compresses the $B_{12}H_{12}^{2-}$ anion, thereby reducing the thermal oscillations of the boron atom. However, this force is counterbalanced by the repulsion forces occurring as the $B_{12}H_{12}^{2-}$ anions approach each other. The stress of the crystalline structure occurring as a result of this counteracting forces may account for the relative difficulty of the synthesis of potassium dodecaborate-bromide (and rubidium dodecaborate chloride) and for the absence of the dodecaborate-fluorides.

On the other hand, an increase in the radii of the cation and halide-ion shall lead to a decrease in the repulsion forces between the $B_{12}H_{12}^{2-}$ anions, and, hence, to the strengthening of the structure. As the radii of the cations and halide-ions increase further the voids assigned to $B_{12}H_{12}^{2-}$ become so large that formation of double salts becomes difficult or even impossible at all.

Hence, the optimum ratio exists between the sizes of the cations and halide-ions which is required for the formation of the most stable double salt on the basis of the dodecaborate of the same element. It is this optimum ratio that can explain the low solubility of bromides in the $Rb_2B_{12}H_{12} \lozenge RbX$ and $Cs_2B_{12}H_{12} \lozenge CsX$ series as compared to the corresponding chlorides and iodides. This can be demonstrated most vividly by comparing the solubility (% by mass) - versus - $\sqrt{X^-}$ Å and - module $[a-(d_{M^+} + d_{X^-})]\lozenge 10^2(\text{Å})- \sqrt{X^-}$ - (Å) graphs where is the cell parameter.

The ideal example in this respect is $Cs_2B_{12}H_{12} \lozenge CsBr$ for which the module $[a-(d_{cs^+} + d_{Br^-})]$ is minimum having the minimum solubility in water.

4. POLYHEDRAL ANIONS $B_N H_N$ AS LIGANDS

The polyhedral boron hydride anions are relatively new and unusual type of ligands in coordination compounds of transition metals. Their peculiarity lies primarily in the fact that the very term ligand presupposes the presence in their composition of the anions capable of donating the electron pair for bonding with the central atom, whereas the polyhedral anions $B_n H_n^{2-}$ are the electrondeficient structures. From this viewpoint there is no similarity between them and such classical ligands as the aromatic systems of benzene, cyclopentadienyl and even ethylene, though the $B_n H_n^{2-}$ structure contains the delocalized electron system. Rather large geometrical dimensions of these anions and the presence of the considerable charge account for their weak coordination capacity with respect to the transition metals.

The most characteristic property of the $B_n H_n^{2-}$ systems in their reactions with transition metal compounds is the insertion of the atom of the metal into the polyhedral system with the formation of metal boranes or, for reactions with carboranes, metal carboranes.

The chemistry of these compounds has been well developed by such researchers as Hottorng, Grimes, Greenwood, et al. [1-5]. The following compounds can be taken as examples. The interaction of $[(h^5\text{-}C_5H_5)Ni(CO)_2]$ with $(R_4N)_2 B_{10}H_{10}$ in the mixed acetonitrile-dimethoxyethane solution yields 45% of $[(h^5\text{-}C_5H_5)_2Ni_2B_{10}H_{10}]$ based on the icosahedron Ni_2B_{10}, with the both nickel atoms being bonded together by the bond nickel-nickel [ ]. Known is the similar compound with cobalt and the mixed complex $(h^5\text{-}C_5H_5)NiCO B_{10}H_{10}$ [88, 89]. Or, for instance, in the reaction between $(R_4N)_2 B_{11}H_{11}$ and nickelocene in the boiling 1.2-dimethoxyethane the compound $(R_4N) [(h^5\text{-}C_5H_5) NiB_{11}H_{11})]$ [88, 89] was obtained, or the 7-apical metal borane $R_4N [(h^5\text{-}C_5H_5)NiB_6H_6]$ was first obtained from $(R_4N)_2B_6H_6$ and nickelocene [90]. The compound $[(CH_3)_4N](h^5\text{-}C_5H_5CoCB_8H_9)]$ [91] or 10-apical closo-metal borane $(h^5 \lozenge C_5H_5)Ni_2CB_7H_8$ obtained by the reaction of nickelocene with 4-CB_8H_9 may serve as examples of numerous series of metal carboranes. The nickel atom in it occupies two neighboring positions in one of the square bases of the structure, (the length of the Ni-Ni bond is 2.618Å), and the carbon atom is located in the center of the opposite square base. The peculiar feature of this structure is that this is a first representative of the stable polyhedral metal carboranes containing the unpaired electron in its skeleton. Just like the CH group, the group C_5H_5Ni introduces three AOs and three electrons into the polyhedral system. Thus, the sum of the skeleton electrons is equal to 23 which does not correspond to the known Wade's rule "2n+2."

The EAS data in the range of 77 to 298K and the data of static magnetic susceptibility, (effective magnetic moment is equal to 1.63 mB), confirm these calculations. Probably, the unpaired electron occupies one of the antibonding MOs which results in the loosening of bonds. Thus the length of the B-B bonds in the B(2)-B(4) region is equal to 1.966Å which is considerably larger than in the initial $B_{10}H_{10}^{2-}$ and the Ni-Ni bonds are considerably stronger than in other similar metal boranes with nickel.

As far as the complexes with closed polyhedral anions are concerned, (without insertion of metal atoms in the polyhedral structure), but the researches here are considerably less intensive and this region is to be studied further.

4.1. COMPLEX COMPOUNDS OF PT(II), PD(II) AND CU(I)

As mentioned in section 2 the triphenylphosphine complexes of the Pd(II) and Pt(II) compound $[ML_2Cl_2]$ react in different ways with the salts containing the polyhedral anions $B_nX_n^{2-}$, where X=Cl, Br. However, in all the cases anions do not displace the ligands from the first coordination sphere; even in such complexes as $[Pt(PPh_2)_2CO_3]$ with strong acids $(H_3O)_2B_nX_n$, $B_nX_n^{2-}$ (X=Cl, Br; n=10, 12) and in interaction of $[Pt(PPh_3)_2Cl_2]$ with $Ag_2B_nX_n$ in the mixture of organic solvents at their boiling temperatures, the final products are the earlier-synthesized $[Pt(PPh_3)_4Cl_2] B_nX_n$. However, this reaction has turned out to be highly efficient when $Ag_2B_nX_n$ is substituted with $K_2B_{10}H_{10}$ (n=10, 12). It was earlier known that the reaction between $K_2B_{10}H_{10}$ and cis- $Pt(PPh_3)_2Cl_2$ in chloroform-ethanol solution results in opening of the boric skeleton and insertion there into of the platinum atom with formation of the 11-apical metal borane cis-$[Pt(PPh_3)_2B_{10}H_{11} OEt]$ [93]. The use of silver salt $Ag_2B_{10}H_{10}$ in place of $K_2B_{10}H_{10}$ and substitution of the chloroform-ethanol mixture with methylene aceton-nitrilechloride have resulted in the formation of two compounds trans- $[Pt(PPh_3)_2HCl]$ and cis $[Pt(PPh_3)_2B_{10}H_{10}]$ [94]. The "B^{31}, P, M$_4$, ^{195}Pt NMR- and IR-spectroscopy analysis have unequivocally confirmed the structure of the latter complex compound. The $B_{10}H_{10}^{2-}$ group plays the part of a bidentate ligand in the complex attached to the central atom through the bonds Pt-H-B.

In its spectroscopic characteristics the complex $[Pt(PPh_3)B_{10}H_{10}]$ is very similar to the earlier-synthesized complex copper (I) compound $[Cu(PPh_3)_2B_{10}H_{10}]\lozenge CHCl_3$ [95] for which the decoding of the crystalline structure has confirmed the bidentate attachment through the Pt-H-B bridging bonds.

More detailed NMR-spectroscopy studies of the cis- $[Pt(PPh_3)_2 B_{10}H_{10}]$ have made it possible to establish the presence of two isomers in it due to different ability of the ligand to coordinate the platinum atom caused by the presence in the polyhedral $B_{10}H_{10}^{2-}$ anion of the unequivalent hydrogen atoms - apical and equatorial. In the one case, the coordination of the $B_{10}H_{10}$ goes on through the bonds Pt-H$_{apical}$-B and Pt-H$_{equat}$-B, while in the other case it is realized due to the bonds Pt-H$_{equat}$-B ad P-H$_{equat}$-B.

In chemical and physical-and-chemical properties the isomers are identical.

When the solation of $[Pt(PPh_3)_2B_{10}H_{10}]$ in acetonitrile is subjected to the action of ethyl alcohol the afore-mentioned 11-apical metal borane $[Pt(PPh_3)_2 B_{10}H_{10}\lozenge OC_2H_5]$ is formed with the platinum atom being inserted into the boric skeleton Hence, it can be assumed that the interaction of $K_2B_{10}H_{10}$ with cis- $[Pt(PPh_3)_2Cl_2]$ in the alcohol solution proceeds through the formation of the substituted compound $[Pt(PPh_3)_2B_{10}H_{10}]$.

The complex cis- $[Pd(PPh_3)_2Cl_2]$ readily enters into the same reactions with the salts containing the $B_{10}H_{10}^{2-}$ anion as the platinum complex with formation of $[Pd(PPh_3)_2B_{10}H_{10}]$. However, the obtained compound is very unstable and for 10 to 15 minutes it completely decomposes with the isolation of metal palladium and evolvement of hydrogen.

At the reaction of the cis- $[Pt(PPh_3)_2Cl_2]$ with $Ag_2B_{12}H_{12}$ in the mixture acetonitrile-methylen chloride the red metal-crystalline complex $[Pt(PPh_3)_2B_{12}H_{12}]$ has been synthesized. Just as the case with $Ag_2B_{10}H_{10}$ the trans- $[Pt(PPh_3)_2HCl]$ is formed as a by-product. In the spectral parameters, $[Pt(PPh_3)_2 B_{12}H_{12}]$, is very close to $[Pt(PPh_3)_2B_{10}H_{10}]$, particularly with respect to the presence of the bridging Pt-H-B bond.

The 31_p {'H} NMR-spectrum is essentially a singlet with d=+12,2 with symmetrically-located satellites due to the interaction with $^{195}Pt(J_{P-Pt}=3670$ Hz).

The "B{'H} NMR-spectrum is a singlet with d=-15.2 m.p. with a clearly expressed arm on the side of the weak field due to the bridging Pt-H-B bond.

As expected, unlike cis- $[Pt(PPh_3)_2B_{10}H_{10}]$ the complex cis- $[Pt(PPh_3)_2B_{12}H_{12}]$ exists as one isomer due to the equivalent character of all the B-H bonds in the icosahedron.

It should be noted that in the spectral characteristics the cis- $[Pt(PPh_3)_2B_{12}H_{12}]$ are very close to the complex Cu(I) $[Cu(PPh_3)_2]_2B_{12}H_{12}$ newly-synthesized by us. It is easily formed with a quantitative yield at the reaction of $[Cu(PPh_3)_3Cl]$ and $Ag_2B_{12}H_{12}$ in the mixture with methylene-acetonitrile chloride.

Regarding palladium (II), the reaction of cis- $[Pd(PPh_3)_2 Cl_2]$ with $Ag_2B_{12}H_{12}$ almost instantaneously leads to catalytic decomposition of the polyhedral ion $B_{12}H_{12}^{2-}$ up to H_3BO_3 and H_2 and with isolation of metal palladium.

Thus, the polyhedral $B_{12}H_{12}^{2-}$ and $B_{10}H_{10}^{2-}$ anions in phosphine complexes of Cu(I), Pt(II) and Pd(II) act as the bidentate ligands attached to the central atom through bridging bonds Cu-H-B, Pt-H-B and Pd-H-B. Stability of phosphine complexes of the compounds of these metals with $B_{10}H_{10}^{2-}$ and $B_{12}H_{12}^{2-}$ sharply falls proceeding from Cu(I) to Pd(II).

4.2. COMPLEX COMPOUNDS OF PB(II)

The interest focused on the compounds of Pb(II) with polyhedral anions is accounted for by the fact that lead is referred to as nontransition element and therefore is not the typical complex former. Interaction between lead carbonate with $(H_3O)_2 B_nH_n$ (n=10, 12) leads to the formation of hydrates of $PbB_nH_n \lozenge XH_2O$ whose dehydration yield, rather hydroscopic powder, containing together with the anion $B_nH_n^{2-}$ and the cation of lead the hydroxyl groups. In this respect it is similar to the dehydrated salts of light alkali metals where the bonds between the cations and polyhedral anions is of a predominantly ionic character.

In the presence of such ligands as 2,2' bipyridyl (Bipy), 1,10-phenanthroline and antipirine (1-phenyl-2,3 dimethylpurazalone-5) compounds the difficult solubility in water of the $PbLB_nH_n$ composition for Bipy and Phen and $PbL_2B_nH_n$ for antipirine are precipitated. These compounds are stable in air and are essentially yellow-green fine-crystalline structures soluble in DMFA, DMSO and nitromethane [96]. The composition of final products does not depend on the initial ratios between the lead salts, ligand and polyhedral anion. The thermal stability of the formed compounds is approximately the same. The specimens are stable up to 340°C, following the intensive exothermal reaction begins with the loss of mass and at 400°C the process of thermoxidative destruction occurs with the destruction of the boron skeleton.

The complex $Pb(Ant)_2B_{10}H_{10}$ loses all the antipirine at as low temperature as 110°C and at further heating within 265 to 380°C the thermooxidative destruction of the lead decaborate occurs.

The IR-spectroscopy data points to the presence of the absorption strips characteristic for the polyhedral anion, with $B_{10}H_{10}^{2-}$ exhibiting the splitting and shift of the absorption strips of the end B-H bonds uncharacteristic for it - the splitting into 5 to 6 components is observed for both apical and equatorial which suggests strong deformation and probably the formation of the bond between the central atom and the decarbonate anion.

For the Pb Bipy $B_{10}H_{10}$ complex complete X-ray crystal structure analysis has been performed - this compound crystallizes in the monoclinic crystal lattice with the parameters a=11.713(2); b=12.172(2); c=11.905(2)Å, Z=4, 3-dimface P_2/n [96].

The structure is formed from complex cations Pb Bipy^{2+} and anions $B_{10}H_{10}^{2-}$. If every $B_{10}H_{10}^{2-}$ anion and Bipy is regarded as a separate element of the structure, then the lead atom is surrounded by three anions and one ligand arranged following a distorted tetrahedron pattern. With the molecule Bipy the lead is bonded through the bond Pb-N with distances 2.500 (5) and 2.470(6)Å, with two anions-through the trigonal faces at the primary B(I) and B(10) with distances from Pb-B from 3.04 to 3.15Å and with the third anion $B_{10}H_{10}^{2-}$ through the edge B(1)-B(2) with distances Pb-B within 3.37 to 3.58Å.

Since the angles for the bonds through the edge are equal to 85 to 99° it can be suggested that the presence of the direct Pb-B bond observed for the first time of the compounds of this type. The angle of bond PbH (B(2)B(2)) is equal to 125 degrees, which probably corresponds to the bond Ph-H-B Pb9B92) 3.37(I); Pb-H (B(2)) 2.79 (8); B(2)-H(B(2)) 1.08Å. With the ligand, the lead atom forms the five-member ring in the form of an envelope. At one corner there is located the lead atom spaced 0.39Å from the plane of the ligand. In this case, the metal ring becomes bent across the N-N line with the dihedral angle of 10.6 degrees. The ligand itself is nearly flat, with the angle between the planes of two piridyl rings equal to 2.5 degrees.

The $B_{10}H_{10}$ anion is of the conventional shape of a two-cap tetrahedral antiprism. The average distances B-B$_{polar.}$ 1.698 (4) B-B$_{trop.}$ 1.827 (4), B-B$_{equator}$ 1.809 (4)Å, where polar is the distance from the "cap" boron atom to the boron atoms of the nearest square base, trop - are the distances inside the base, equator - the distances between the boron atoms of different bases. These distances are close to the similar data for the 10-apical closo-polyhedral borane complexes. The general pattern of the structure can be described as anion layers located approximately in the planes (101). The distance between the layers is about 9Å. These layers are bonded together by the cations, the planes of which are approximately perpendicular to the planes of the layers. Here, it is as if the ligands from partially turned-over pairs at distances of about 3.8 to 4.0Å.

4.3. COMPLEX ANION $CuB_{10}H_{10}$

The salts of Cu, $Cu(ClO_4)$ or $Cu(OH)_2$ in particular, rather readily interact with the salts of the dodecohydro-closo-dodecaboric acid, for instance with $Na_2B_{10}H_{10}$ or directly with the acid (for the case with $Cu(OH)_2$) with the formation of the salt of monovalent copper - $Cu_2B_{10}H_{10}$ [97] difficultly soluble in water. The peculiar feature of this compound is basically the covalent character of the bond between the cation Cu(I) and the polyhedral anion. In the crystal, the atoms are located nearer than the centers of the opposite deltahedrons of the polyhedron in such a way that four copper atoms (I) are coordinated around the anion and the entire structure can be regarded as a three-dimensional covalently-bonded frame $(Cu_2B_{10}H_{10})_n$. The distances Cu-B lie within 2.14 to 2.33Å, (i.e. to the distance anticipated for the 3-center bond B-Cu-B (2.13Å). In other words, the multinuclear complex is formed where $B_{10}H_{10}^{2-}$ plays the part of ligands. In the other complex $[Cu(PPh_3)_2]_2B_{10}H_{10}$ $CHCl_3$ the polyhedral anions $B_{10}H_{10}^{2-}$ are already attached to the central atom through the bridging Cu-H-B bonds [95]. Each polyhedral anion bonds two crystallographically-unequivalent copper atoms in such a way that they transform into each other by the symmetry operation S_8. The average distance is greater than in Cu (H_{equiv}=2.08Å) and 1.96Å, respectively). The difference in the lengths of the bonds is also accounted for by the three-center Cu-H-B interaction. On the basis of the structural data and vibration spectra Lipscomb et al. [97] have come to the conclusion that in these cases the three-center B-Cu-B interaction most likely takes place. Such a model explains the disappearance of the absorption strips at 1015 and 1070 cm^{-1} in the IR-spectra. It should be noted that $[Cu(PPh)_2]_2B_{10}H_{10}$ is readily transformed into $Cu_2B_{10}H_{10}$ under, for instance, action of diborane.

The tendency of transforming Cu(II) into Cu(I) in the presence of the polyhedral anions is observed in the reactions of various copper (II) salts, in which case the authors have observed the other type of the final products. If the reactions between $CuCl_2$ and $K_2B_{10}H_{10}$ (or $Cs_2B_{10}H_{10}$) are conducted at elevated temperatures, then after 4 hours the pale-blue crystals $KCuB_{10}H_{10}$ (or $CsCuB_{10}H_{10}$) are precipitated. These compounds are difficultly-soluble in water, and readily soluble in acetonitrile, acetone, nitromethane, dimethylformamide and dimethylsulphoxide.

In the same way the compounds of the composition $(R_nNH_{4-n})CuB_{10}H_{10}$ have been obtained, where n=2.4; $R=CH_3, C_2H_5, C_4H_9$, etc. The initial compounds with the polyhedral anions are $[R_nNH_{4-n}]_2B_{10}H_{10}$.

It should be noted that the attempts to perform in the compounds the direct exchange of the cations with the anion $[CuB_{10}H_{10}]^-$ have always resulted in the stable $Cu_2B_{10}H_{10}$.

The presence of the anion $[CuB_{10}H_{10}]^-$ in all the afore-mentioned compounds has been proved by the combination of the physical and physical-chemical properties. All the compounds have turned out to be diamagnetic, i.e. containing only Cu(I).

Inclusion of Cu(I) in the compound strongly affects the state of $B_{10}H_{10}^{2-}$ - in the IR spectra there is observed strong splitting of the valet oscillation of B-H_B and B-H bonds and the appearance of new broad strips at 2265 and 2355 cm^{-1}, associated with the bridging Cu-H-B bonds. For the compounds with the organic cations the bridging Cu-H-B bonds have the maximum at 2140-2160 cm^{-1}. However, in the compounds of this type there is observed the strongly-pronounced effect of the organic cation on the polyhedral anion probably because of the formation of the hydrogen bonds. In other words, the polyhedral anion in these compounds is under the strong effect of both the copper (I) atom and the organic cation. This results, for instance in a considerable decrease in the thermal stability of the compounds with $[CuB_{10}H_{10}]^-$ as compared with similar compounds with $B_{10}H_{10}^{2-}$. For instance, if the thermal decomposition $Cu_2B_{10}H_{10}$ begins at temperatures higher than 600°C then for $KCuB_{10}H_{10}$ and $CsCuB_{10}H_{10}$ if takes place at as low temperatures of 290-300°C. For complexes with organic cations this difference is smaller but, all the same, it lies within 80 to 100°C.

The bond of copper with the polyhedral anions is insufficiently strong to ensure stability of the complex anion in the solution, since at the solution the stable compound $Cu_2B_{10}H_{10}$ is rapidly formed.

The failure of attempts to conduct the exchange reactions for substitution of the cations in these compounds and the results of the NMR-spectroscopy analysis of the solutions of these compounds makes it possible to suggest that these compounds exist in the form of polymer-similar structures which are easily destroyed at solution.

CONCLUSION

The chemistry of cluster boron compounds is wide and diverse. In itself it comprises part of the coordination chemistry fruitful in theoretical aspect and very interesting in the experimental respect. As has been noted, its many fundamental aspects lie in-between inorganic (particularly, coordination), element-inorganic and organic chemistry. From this point of view there are still many problems in it whose solution requires the effort of researchers and may yield important results not just for the chemistry of the boron compounds.

The chemistry of the polyhedral boron hydride compounds is an inexhaustible source of new interesting ligands in the coordination chemistry of not only typical complex formers, transition metals, but, as research has shown, for non-transition metals as well, despite the weak coordination capacity of $B_{10}H_{10}^{2-}$.

As the outer-sphere anions in the coordination compounds the polyhedral anions $B_nH_n^{2-}$, $B_nH_{n+1}^-$ and $CB_{n-1}-H_n^-$ noticeably affect the composition of the first coordination sphere and its thermal and kinetic stability, the way in which the ligands are coordinated to the central atom in cases where the ligand contains no less than two atoms capable of forming a chemical bond with the central atom.

This interesting field of coordination chemistry is still at the initial stage of development. Its problems are still being tackled by a relatively narrow group of specialists in the growth of chemistry of boron-hydride compounds and their derivatives, but the authors believe that in the near future these compound shall be easily accessible for complex-chemists.

REFERENCES

1.   Grimes, R.N., "Carboranes," Academic Press, NY, 1970.
2.   "Comprehensive Organometallic Chemistry," eds Wilkinson, G., Stone, F.G.A. and Abel, E., Pergamon Press, Oxford, 1982.
3.   "Metal Interaction with Boron clusters," ed. Grimes, R.N. Plenum Press, Z., N.Y. 1982.
4.   Greenwood, N.N., *J. Pure Appl. Chem.*, 1983, v. 55, p. 77.
5.   Greenwood, N.N., *Chem. Soc. Rev.*, 1984, v. 13, p. 353.
6.   King, R.B., Rouvray, D.H., *J. Amer. Chim. Soc.*, 1977, v. 99, p. 7834.
7.   King, R.B., Rouvray, D.H., *Theor. Chim. Acts.*, 1978, v. 48, p. 207.

8. Agafonov, A.V., Solntsev, K.A., Kuznetsov, N.T., *Coord. Chem.* 1980, v. 6, p. 252.
9. Lipscomb, W.N., "Boron Hydrides," Benjamin, W.A., N.Y. 1963.
10. Muetterties, E.L. and Knoth, W.H., "Polyhedral Boranes," Dekker, N.Y. 1968.
11. "Boron Hydride Chemistry," ed. Muetterties, E.L., Academic Press, N.Y. 1975.
12. "Boron Chemistry-4," ed. Pavry, R.W. and Kodama, G., Pergamon Press, Oxford, 1980.
13. Longuet-Higgins, H.C., *J. Chem. Phys.*, 1949, v. 46, p. 268.
14. Longuet-Higgins, H.C., *J. Roy. Inst. Chem.*, 1953, v. 77, p. 179.
15. Longuet-Higgins, H.C. and de V. Roberts M., Proc. Roy, *Soc.* (London), 1955, v. A230, p. 110.
16. Wade, K., *Adv. Inorg. Chem. Radiochem.*, 1976, v. 18, p. 1.
17. "Transition Metal Clusters," ed. Johnson, B.F.G., Chichester, Wiley, 1980.
18. Williams, R.E., *Inorg. Chem.* 1971, v. 10, p. 210.
19. Williams, R.E., *Adv. Inorg. Chem. Radiochem.*, 1976, v. 18, p. 67.
20. Mason, R., Thomas, K.M. and Mingos, D.M. J *Amer. Chem. Soc.*, 1973, v. 95, p. 3802.
21. Mingos, D.M.P., *Nature (L) Phys. Sci.*, 1972, v. 236, p. 99.
22. Mingos, D.M.P., *Adv. Organometal. Chem.*, 1977, v. 5, p. 1.
22a. Mingos, D.M.P., Foryth, M.I., *J. Chem. Soc. Dalton Trans.*, 1977, p. 61.
23. Rudolph, R.W., *Acc. Chem. Res.*, 1976, v. 9, p. 446.
23a. Rudolph R.W. and Pretzer, W.R., *Inorg. Chem.*, 1972, v. 11, p. 1974.
24. Zhu Longgen, Feng Xinghong, *J. Nanjiing University*, 1985, v. 21, p. 277.
25. Au-chin Tang and Qian-Shu Li Intern., *J. Quant. Chem.*, 1986, v. 29, p. 579.
26. Muetterties, E.L., Balthis, J.H., China, Y.T., Knoth, W.H. and Miller, H.C., *Inorg. Chem.*, 1964, v. 3, p. 444.
27. Knoth, W.H., Miller, H.C., Saner, J.C., Balthis, J.H., China, Y.T. and Muetterties, E.L., *Inorg. Chem.*, 1964, v. 3, p. 159.
28. Kuznetsov, N.T., Kulikova, L.N., Faerman, V.I., Proc. Russian Series, "Inorg. materials," 1976, v. 12, p. 1212.
29. Kuznetsov, N.T., Kulikova, L.N., Zhukov, S.T., *J. Inorg. Chem.*, 1976, v. 21, p. 96.
30. Kulikova, L.N., Kuznetsov, N.T., Lappo, I.V., Zhukov, S.T., *J. Inorg. Chem.* 1976, v. 21, p. 933.
31. Kuznetsov, N.T., Kulikova, L.N. Proc. Russian series "Inorg. Chem.," 1980, v. 16, p. 1403.
32. Zhukova, N.A., Kuznetsov, N.T., Solntsev, K.A., Ustunyuk, Yu.A., Grishin, Yu.V., *J. Inorg. Chem.*, 1980, v. 25, p. 690.
33. Zhukova, N.A., Kuznetsov, N.T., Solntsev, K.A., *J. Inorg. Chem.*, 1980, b. 25, p. 925.
34. Zhukova, N.A., Kuznetsov, N.T., Solntsev, K.A. *J. Inorg. Chem.*, 1980, v. 25, p. 2939.

35. Fritze, J., Preetz, W. and Marsmann, H.C., Z. Naturfovsch, 1987, v. 426, p. 287.
36. Preetz, W., Henrich, A. and Thesing, J., Z. Naturforsch., 1988, v. 43b, p. 1319.
38. Henrich, A. and Preetz, W., Marsmann, H.C., Z. Naturoforsch, 1988, v. 43b, p. 647.
39. Preetz, W., Fritze, J., Z. Naturoforsch, 1984, v. 39b, p. 1472.
40. Preetz, W., Fritze, J., Z. Naturoforsch, 1987, v. 42b, p. 282.
41. Preetz, W., Fritze, J., Z. Naturoforsch, 1987, v. 42b, p. 293.
42. Agafonov, A.V., Butman, L.R., Solntsev, K.A., Vinokurov, A.A., Zhukova, N.A., Kuznetsov, N.T., J. Inorg. Chem., 1982, v. 27, p. 63.
43. Yakushev, A.B., Sivaev, I.B., Solntsev, N.T., Coord. Chem., 1990, v. 16.
44. Yakushev, A.B., Sivaev, I.B., Solntsev, N.T., Coord. Chem., 1990, v. 16.
45. Wade, K., "Electron Deficint Compounds," Nelson, London 1971.
46. Evans, J., J. Chem. Soc., Dalton Trans. 1978, p. 25.
47. Brint, P., Healy, E.E., Spalding, T.R. and Whelan, T., J. Chem. Soc., Dalton Trans. 1981, p. 2515.
48. Whelan, T., Brint, P., J. Chem. Soc., Dalton Trans. 1983, p. 975.
49. Cavanaugh, M.A., Fehlner, T.P., Stramel, O'Neill, M.E. and Wode, K. Polyhedron, 1985, v. 4, p. 687.
50. Housecroft, C.E., Snaith, R., Moss, K., Muvey, R.E., O'Neill, M.E. and Wde, K., Polyhedron, 1985, v. 4, p. 1875.
51. Dekock, R.L., Jasperse, C.P., Inorg. Chem., 1983, v. 22, p. 3843.
52. Mebel, A.M., Charkin, O.P., Kuznetsov, I.Yu., Solntsev, K.A., Kuznetsov, N.T., J. Inorg. Chem., 1988, v. 33, p. 1685.
53. Vinitsky, D.M., Lagun, V.L., Solntsev, K.A., Kuznetsov, N.T., Kuznetsov, I.Yu., Coord. Chem., 1985, v. 11, p. 1504.
54. Kuznetsov, I.Yu., Vinitsky, D.M., Solntsev, K.A., Kuznetsov, N.T., Butman, L.A., Proc. USSR AS, 1985, v. 283, p. 872.
55. Vinitsky, D.M., Solntsev, K.A., Kuznetsov, N.T., Goeva, L.V., J. Inorg. Chem., 1986, v. 31, p. 2326.
56. Privalov, V.I., Tarasov, V.P. Keladze, M.A., Vinitsky, D.M., Solntsev, K.A., Buslaev, Yu.A., Kuznetsov, N.T., J. Inorg. Chem., 1986, v. 31, p. 1113.
57. Mebel, A.M., Charkin, O.P., Solntsev, K.A., Kuznetsov, N.T., J. Inorg. Chem, 1989, v. 34, p. 283.
58. Vinitsky, D.V., Rezvova, T.V., Solntsev, K.A., Kuznetsov N.T., Abstr. IV-All Union meeting on chemistry of inorgan. hydrides Dushanbe, 1987, p. 5.
59. Mebel, A.M., Charkin, O.P., Solntsev, K.A., Kuznetsov, N.T., J. Inorg. Chem., 1988, v. 33, p. 2263.
60. Mustyatsa, V.N., Votinova, N.A., Solntsev, K.A., Kuznetsov, N.T., Publ. USSR AS 1988, v. 201, p. 1396.
61. Gaft, Yu.L., Zakharova, I.A., Kuznetsov, N.T., J. Inorg. Chem., 1980, v. 25, p. 1308.
62 Gaft, Yu.L., Kuznetsov, N.T., J. Inorg. Chem., 1981, v. 26, p. 1301.

63. Voronkov, M.G., Pukharevich, V.B., Iaykhanskaya, I.L., Ushakova, N.Y., Gaft, Yu.L., Zakharova, I.A., *Inorg. Chim. Acta*, 1983, v. 68, p. 103.
64. Kukina, G.A., Sergienko, V.S., Gaft, Yu.L., Zakharova, I.A., Porai-Koshits, M.A., *Inorg. Chim. Acta*, 1980, v. 45, p. 257.
65. Kayumov, A.G., Solntsev, K.A., Goeva, L.V., Kuznetsov, N.T., Ellert, O.G., *J. Inorg. Chem.*, 1988, v. 33, p. 1936.
66. Kayumov, A.G., Yakushev, A.B., Solntsev, K.A., Goeva, L.V., Kuznetsov, N.T., Ellert, O.G., *J. Inorg. Chem.*, 1988, v. 33, p. 2587.
67. Kayumov, A.G., Solntsev, K.A., Goeva, L.V., Kuznetsov, N.T., *J. Inorg. Chem.*, 1988, v. 33, p. 1201.
68. Kauymov, A.G., Solntsev, K.A., Goeva, L.V., Kuznetsov, N.T., Ellert, O.G. *J. Inorg. Chem.*, 1988, v. 33, p. 1771.
69. Zhaug, Lun, Hu, Peizhi; Gao, Zin. Wuji Huaxue, 1986, v. 2, p. 60.
70. Zhang, Guomin, Jiang, Tengchao, *Zhang Lun Wuji Huaxue*, 1987, v. 3, p. 51.
71. Kuznetsov, N.T., Zemskova, L.A., Goeva, L.V., *Coord. Chem.*., 1981, v. 7, p. 232.
72. Kuznetsov, N.T., Zemskova, L.A., Alikhanova, Z.M., Ippolitov, E.G., *J. Inorg. Chem.*, 1981, v. 26, p. 1331.
73. Kuznetsov, N.T., Zemskova, L.A., *J. Inorg. Chem.*, 1982, v. 27, p. 1320.
74. Kuznetsov, N.T., Zemskova, L.A., Ippolitov, E.G., *J. Inorg. Chem.*, 1981, v. 26, p. 2501.
75. Mikhaylov, Yu.N., Kanishcheva, A.S., Zemskova, L.A., Mistrykov, V.E., Kuznetsov, N.T., Solntsev, K.A., *J. Inorg. Chem.*, 1982, v. 27, p. 2343.
76. Kuznetsov, N.T., Zemskova, L.A., Ippolitov, E.G., *J. Inorg. Chem.*, 1981, v. 26, p. 1862.
77. Kuznetsov, I.Yu., Solntsev, K.A., Kuznetsov, N.T., Mikhaylov, Yu.N., Orlova, A.M., Alikhanova, Z.M., Sergeev, A.N., *Coord. Chem.*, 1986, v. 12, p. 1387.
78. Kuznetsov, N.T., Klimchuk, G.S., *J. Inorg. Chem.*, 1971, v. 16, p. 1166.
79. Kuznetsov, N.T., Klimchuk, G.S., Kanaeva, O.A., *J. Inorg. Chem.*, 1975, v. 20, p. 2557.
80. USA Patent 3, 411, 890 (1968).
81. Kanaeva, O.A., Kuznetsov, N.T., Sosnovskaya, O.O., Goeva, L.V., *J. Inorg. Chem.*, 1980, v. 25, p. 2380.
82. Kanaeva, O.A., Klimchuk, G.S., Solntsev, K.A., *J. Inorg. Chem.*, 1987, v. 32, p. 803.
84. Kanaeva, O.A., Kuznetsov, N.T., Sosnovskaya, O.O., *J. Inorg. Chem.*, 1978, v. 23, p. 3347.
85. Kanaeva, O.A., Kuznetsov, N.T., Sosnovskaya, O.O., *J. Inorg. Chem.*, 1981, v. 26, p. 1153.
86. Solntsev, K.A., Kuznetsov, N.T., Trunov, V.K., Karpinsky, O.G., Klimchuk, G.S., Uspenskaya, S.I., Oboznenko, Yu.V., *J. Inorg. Chem.*, 1977, v. 22.

87. Kuznetsov, N.T., Klimchuk, G.S., Kanaeva, O.A., Solntsev, K.A., *J. Inorg. Chem.*, 1976, v. 21, p. 927.
88. Sullivan, B.P., Leyden, R.N., Hawthorne, M.F., *J. Amer. Chem. Soc.*, 1975, v. 97, p. 455.
89. Leyden, R.N., Sullivan, B.P., Baker, R.T., Hawthorne, M.F., *J. Amer. Chem.,Soc.* 1978, v. 100, p. 3758.
90. Vinitsky, D.M., Lagun, V.L., Solntsev, K.A., Kuznetsov, N.T., Karushkin, K.N., Yanoushek, Z., Bashe, K., Shtibr, B., *J. Inorg. Chem.*, 1984, v. 29, p. 1714.
91. Solntsev, K.A., Butman, L.A., Kuznetsov, N.T., Kuznetsov, N.T., Shtibr, B., Yanovshek, Z., Bashe, K., *Coord. Chem.*,1983, v. 9, p. 993.
92. Solntsev, K.A., Butman, L.A., Kuznetsov, N.T., Kuznetsov, N.T., Shtibr, B., Yanoushek, Z., Bashe, K., *Coord. Chem.*, 1984, v. 10, p. 1132.
93. Paxson, T.E., Hawthorne, M.F., *J. Inorg. Chem.*, 1975, v. 14, p. 1604.
94. Gaft, Yu.L., Ustynyuk, Yu.A., Borisenko, A.A., Kuznetsov, N.T., *J. Inorg. Chem.*, 1983, v. 28, p. 2234.
95. Paxson, T.E., Hawthorne, M.F., Brown, L.D. and Lipscomb, W.N., *Inorg. Chem.* 1974, v. 13, p. 2772.
96. Malinina, E.A., Solntsev, K.A., Butman, L.A., Kuznetsov, N.T., *Coord. Chem.*, 1989, v. 15, p. 1039.
97. Dobrott, R.D., Lipscomb, W.N., *J. Chem. Phys.*, 1962, v. 37, p. 1779.

HIGH TEMPERATURE
COORDINATION CHEMISTRY

S.V. Volkov
Kiev, IONKh, Academy of Sciences
Ukraine

Abstract—This summary of the author's and other scientists' works outlines the fundamental concepts of high-temperature coordination chemistry as a new scientific trend: objectives, concepts and terms.

This book discusses the questions of the chemistry of coordination compounds: in ionic melts (complex-cluster structure model; quazi-nuclear heteronuclear complexes; quantum chemistry and spectroscopy of radiative and non-radiative transitions in complex melts, etc.); in unhydrous high-boiling solvents (problem of simultaneous exhibition of properties of solvent-ligand-reagent; heteronuclear and multinuclear complexes; solvatometallurgical aspects, etc.); in gas phase (quantum chemistry, spectroscopy and structure of complexes; problems and criteria of volatility of complexes; gas-transport, home- and hetero-phase processes with their participation); in laser chemistry (for compounds in gas phase; complexes in melts and polymerizing and solid phases).

I. INTRODUCTION.

OBJECTIVES, SUBJECT MATTER AND CONCEPTS

High-temperature coordination chemistry is a new branch of inorganic and coordination chemistry, embracing such specific aspects as coordination compounds, such as in ionic melts; anhydrous high-boiling liquids; the gaseous phase; and in some cases under laser and transient plasma conditions, etc., (i.e. under such conditions where the external equilibrium or transient internal temperatures, which become commensurable with or approach the energies of bonding between and inside the complexes [1-6], cannot be neglected).

High-temperature coordination chemistry is attracting the attention of scientists and engineers. The scientific importance of this branch of chemistry is found in the unique phase states, (in some cases), of such compounds; by the possibility of obtaining unusual ratios between concentrations of metals and ligands or untypical valency of the central ions of metals; the problems of stereochemical nonrigidity and the highly dynamic character of some configurations of complexes or by the problems of rigid character of these configurations distorted by crystal fields in other media; by superimposure of the temperature-excited states of particles, etc. not to mention various kinds of chemical interactions indifferent mediums and conditions. Table 1 compares some of the major specific features of coordination compounds of chemistry in the various homogeneous systems depending on temperature and, subsequently, on phase states.

Engineers pay a considerable amount of attention to this branch of chemistry because of the possibility of utilizing these compounds in novel engineering concepts and designs, taking into account the dire need for saving raw materials, fuel, power and material resources.

Table 1. Peculiarities of chemistry of coordination compounds in homogeneous systems depending on the phase state (and temperature)*

Nos in crystal	In molten salt	In gas
1. Uniform unchanged composition of complex	Set of complexes and possibility of realization of conditions CM>CL	Uniform unchanged composition of complex
2. Absence of medium solvent	Presence of medium solvent. Notions solvent-ligand come closer together in binary systems MLn-AlkL	Absence of the medium solvent
3. Presence of crystal lattice-non-isolated complexes (+intermolecular forces)	Presence of "quasi lattice" - non-isolated complexes (+interaction with outersphere cations Alk^+)	Absence of crystal lattice-isolated complexes (intramolecular forces)
4. Problem of rigid character of configuration MLm distorted by crystal field	Problem of distorted character of configuration ML_m and dynamic character of configuration $Alkn[ML_m]$	Problem of stereochemical non-rigidity and dynamic character of configuration ML_m
5. Neglecting excited oscillatory states	Superimposure of temperature-excited oscillatory states	Superimposure of temperature-excited spinning-oscillatory states
increase in concentration of individual complex		

*M - complex-forming metal; L - ligand, Alk+ - outer-sphere cation
(normally of alkali metal; m - number of coordinated ligands

Coupled with the limitations encountered in finding new and expanding existing water- and power-consuming processes, the theoretical studies and practical researches aimed at developing the water-free small-energy-consumption technologies where the desired rates of chemical, metallurgical and other processes obtained at minimum consumption of energy and resources are promising. This aim can be achieved through the use of both chemical, (complex formation, solvato-complex formation), and physical, (temperature, laser sources, plasma), principles of stimulating and controlling chemical processes. Hence, the broad opportunities that can be offered by the researches into the complexes in unhydrous melts, high-boiling solvents and the gaseous phase with the use of lasers and plasma; can now be seen (i.e. the research in the field of high-temperature coordination chemistry).

On the other hand, much interest now focused on the high temperature coordination chemistry arises from the fact that it offers prospects of success in solving a number of the global energy consumption problems and in providing the world with ecologically-clean energy [3,5,6].

Presently, only 20% of the world's energy resources are being consumed as electrical power. The remaining 80% are being used as fuels and non-electrical heat; 30% as low-grade domestic and utility heat; 30% as high-grade heat and energy-conveying media in chemical, metallurgial and other industries, and 20% as transport fuels. When added to hydraulic and thermal power plants, nuclear, (and in the future, thermonuclear), power plants will definitely be able to provide the world with the necessary *amount* of energy, but because of generating only electrical power and heat (in approximately equal proportions), they will not be able to satisfy the demands for the *quality* of energy. Besides, the heat now generated by power plants is low-temperature and low-grade.

What are the ways for eliminating such a qualitative unbalance between electrical and heat power and between the desired high-grade and the generated low-grade heat?

Some ways which are now visualized are as follows:

- transformation of non-electrical technologies into electrical-, plasma- and laser-engineering technologies and the solution of the problem of electrical power supply of transport with self-contained powerplants;
- conversion of the prime energy of nuclear power plants, (mainly heat), into the secondary power resources, for instance, hydrogen can be used in chemical, metallurgical and other processes and as alternative fuel for transport with self-contained powerplants;
- development of high-temperature, safe nuclear power plants to be operated in conjunction with metallurgical, chemical and petrochemical enterprises, and the combined utilization of electrical power + secondary power resources + high-grade heat.

In all of these cases, the high-temperature coordination chemistry may play an important part. However, primary consideration should be given to the following facts. The modern thermal and nuclear power plants, (even without the effect produced by other industries), evolve huge amounts of heat. Taking into account the fact that demands in power double every 10 to 15 years, the heat "contamination" aggravated by the "greenhouse" effect may become catastrophic for the earth unless a reasonable way of utilizing this heat is found. Even though the use of heat for agricultural and utility purposes may be regarded as the first "step" in heat utilization, it cannot solve the problem as a whole, because this way of utilizing heat means mere dissipation.

The most promising way of "cleaning" the earth from heat contamination is to direct it into *endothermal* chemical and other processes. Since heat wastes and discharges are, as a rule, low-grade and low-temperature, their utilization will require preparation of the reagents with weaker individual chemical bonds, (i.e. the reagants which would become transformed at these temperatures).

Therefore, the authors regard the intensive development of the high-temperature coordination chemistry [1-6] as one of the ways to solve this problem. Using this way would make it possible to either realize part of the process under "low-temperature" conditions allowing the utilization of the evolved heat, (and consequently neither heating the atmosphere nor consuming the nonrecoverble prime energy resources), or to produce energy-accumulating substances. In addition, these processes can be performed in water-free and close-loop cycles, i.e. the problems of power engineering, ecology and economy can be solved at the same time. Thus, the afore-mentioned information is strong evidence in favor of the need for narrowing the gap between technology and power engineering, (i.e. for developing the processes utilizing huge amounts of low-grade heat or synthesizing them with the use of energy accumulating substances.

The *principle* of high-temperature coordination chemistry consists of the following; Many chemical and metallurgical processes require high temperatures because of the high chemical bonding energy of the simple compounds used in the industrial process (for metallurgy: oxides, metal sulphides, etc.). In some cases the simple compounds may be transformed into coordination compounds, as a result of which the energy of each separate bond, owing to a large number of the formed bonds, decreases to the level $(G_{bond} = \dfrac{G_{complex}}{coordination\ number})$, thus allowing for the thermal, chemical, electrochemical, photochemical and laser-chemical transformation of molecules and permitting the processes to be conducted under considerably miler conditions with utilization of low-grade heat. But this is only a *thermodynamical* principle used in solving the problem.

There are also advantages to the kinetic character. Unlike the heterogeneous systems, the homogeneous systems, (coordination compounds in liquids and gases), experience much less kinetic difficulties due to the removal of diffusion barriers. And since the diffusion processes are inconsiderably affected by the temperature alone, the rates of chemical reactions increase exponentially with the temperature. Besides, because of the presence of the liquid medium, (rather than vacuum), the activation energies often turn out to be lower than the bond disruption energy.

It is important also to note that it is *the structure* of both the complex itself and the complex with account taken of the surrounding medium that results in more favorable thermodynamic and kinetic parameters of these processes. While the simple compound is being transformed into the complex one, changes in the compound structure are sometimes accompanied with one more positive factor, namely with the conversion from the polymeric and island-like structure, which contributes to the higher volatility of the complex and allows the gas-transport reactions to be realized with the participation of coordination compounds under lower temperatures.

There is an intimate connection between the high-temperature inorganic and laser chemistry. The selective laser excitation of the desired energy levels in molecules, (remember that coordination complexes considerably expand the nomenclature of resonantly-excited molecules), enable promising chemical reactions and processes to be realized at equilibrium room temperatures due to the selective oscillatory or intensive non-equilibrium thermal excitation.

Now, prior to briefly outlining the results of some particular researches conducted at the Institute of General and Inorganic Chemistry of the Academy of Sciences of the Ukraine SSR the authors will try to formulate the family of compulsory attributes of the coordination compound in any medium, (Table 2), rather than in the solution alone, as this is done in [7].

Apart from the high-temperature synthesis, the studies of complexes in melts, high-boiling liquids and gases include studies at high temperatures in various phase states with the use of such advanced research methods as electronic, infrared, Raman scattering spectroscopy, gas-liquid chromatography, chromate-mass-spectrometry, holographic and other high-temperature techniques (Table 3).

Table 2. General attributes of the coordination compound
in homogeneous (one-phase) systems [8]

in crystal in melt in solution in gas	
Attribute I	coordination phenomenon: i.e unchanged geometrical arrangement of anions of one kind around the other: complex-forming metal
Attribute II	complex composition which manifests itself in a discrepancy between the formal degree of oxidation of complex-forming metal and the coordination number and consists in marked rearrangement of electron shells of interacting particles with formation of partially covalent bond;
	(Attribute III: measurable incomplete
	incomplete heterolytical-type dissociation of compound)
	lifetime which exceeds the duration of contact of un-coordinated particles and is essentially a constant value characteristic for any thermonuclear and kinetic property
Attribute III	
increasing concentration of individual complex	

The proper high-temperature coordination synthesis is performed at temperatures ranging from 100 to 350°C, depending on the medium, conditions, kind of metal, salt, oxide, sulphide, etc. and corresponding high-temperature medium ligands. The subsequent transformation of complexes into desired compounds and substances may be realized following one of the above-mentioned methods: thermal, thermochemical, electrochemical, laser-chemical, etc. in various media and conditions and practically within the same temperature interval.

Table 3. Methods used for studying the structure of coordination compounds in melts and vapors

Method	Approximate time scale, s	Application	
		in melts	in vapors
Electron diffractometry	10^{-2}	– –	++
Neuronography	10^{-18}		– –
Radiography	10^{-18}	+ –	– –
Ultraviolet and visible-spectrum radiography	10^{-15} to	++	
Infrared (IR) and Raman scattering (RS) spectroscopy	10^{-14}	++	
	10^{-13} to		
Microwave spectroscopy	10^{-11}	– –	++
	10^{-10} to		
EPR	10^{-8}	+ –	– –
	10^{-9} to		
NMR	10^{-4}	+ –	– –
Mossbauer spectroscopy	10^{-9} to 10^{-1}	– –	– –
Studies in molecular beams	10^{-7}	– –	++
Symbols	10^{-6}		
used to advantage – ++		Duration of contact of free ions in melts: 10^{-10} to $10^{-12}{}_5$	Lifetime of fundamental state in vapors for complexes $Mo(CN)^{4-}{}_8$, $ReH^{2-}{}_9$, etc., 10^{-3} to $10^{-12}{}_5$
there are a number of – – +			
studies			
has recently been			
brought in use –			
not used – – –			

II. CHEMISTRY OF COORDINATION COMPOUNDS IN IONIC METALS

The systems of molten salts, binary with common anion MX-AlkX, (more seldom with common cation Mx-MY), more complex - interrelated tertiary of the MX-AlkY type, etc.[1] , due to the unequal value of their components, (in the explicit form - ion moments of cations M and Alk=charge/radius, and in the implicit form - their electronic structure and other parameters), are characterized by the tendency towards strong interparticle interaction, which in the case of considerably differing parameters of the partners (for instance, for the alkali metal halide systems - (AX-A) Alk=A and d or f-metals (MX) are transformed into a formation of complexes.

Unlike the processes of complex formation in hydrous and anhydrous media, the complex formation in molten salts is distinguished by a peculiar feature found in the absence of generally-accepted solvents due to which the "background" ions, (usually, cations Alk^+), of the molten medium are located on the outer spheres and are not separated from complexes, (normally, anions of the MX_m^{n-} kind, m is the coordination number), by the solvate shells of solvents. As a result, being located on the second coordination sphere of the complex, these ions enter into a considerably intensive, partially covalent, interaction with it, with the resultant formation of quasicluster groups, both heteronuclear $Alk_n[MX_m]$ and homonuclear $Mp[MX_m]$: the former ones are binary, tertiary, etc. molten salts systems, the latter ones, in the proper molten complex-forming metal salt or in the concentration region of the compound rich with this salt.

Such a quasicluster approach to the chemistry of coordination compounds in molten salts makes it possible to discuss the problem of formation of complexes in them and the problem of forming melts in general, to study the thermodynamics and kinetics of the interaction and to propose a number of quantum-chemical approaches and models [9].

The authors will not discuss more conventional cases for the formation of a complex MX_m^{n-} ion in the molten solvent medium (Solv), for instance Al-kNO$_3$, since, in this case, the process of complex formation in the melt is similar to the same process in any solvent consisting in the displacement of the molecules of the solvent, (Solv, -NO$_3$- and others), from the coordination sphere of solvatocomplex $M(Solv)_m^{n-}$ by the stronger ligand (X=Cl$^-$, Br$^-$, J$^-$, CN$^-$, etc.).

The cases of complex formation in the binary systems of the MX-Alk type, which bear resemblance to the formation of aqua-(solvato-) complexes are of much greater interest, because, in this case, one part of the anions in the melt (X-1) acts as the ligand and the other part as the solvent.

The aim of this section[2] of the paper is to consider the following problems of interest for molten salts:

---

[1] To simplify matters, the formal degrees of oxidation and corresponding stoichiometric coefficients are not discussed.

[2] Preparation of this section was assisted by Babushkina, O.B., Bandur, V.A., Buryak, N.I. and others.

- a study of the presence in such systems of the true, (in classical interpretation), complex ions and the determination of the region of their existence - complex-cluster model;
- a study of the question of the similarity or dissimilarity of these complex ions in melts and commonly-accepted complexes in hydrous and anhydrous media, in order to allow the application of the known theoretical and experimental laws and relationships, (quasibinuclear heteronuclear complex systems), to the former ones (in case of their similarity);
- in the case of their dissimilarity, (because of specific features of molten salts), development for them of such quantum chemistry concepts which would explain the available experimental results and would possess the predicting capability;
- determination of common and specific features of coordination of ions in molten salts which would make it possible to proceed to consideration of the problems of coordinated molecular solvents.

1. COMPLEX-CLUSTER MODEL OF STRUCTURE OF MOLTEN-SALT SYSTEMS WITH COMPLEX FORMATION

Let us consider the structure of the molten salt systems with a MX-AlkX type complex formation, for which purpose two characteristic regions of their composition shall be discussed.

In the case of diluted solutions of M^{Z+} ions with considerable excess amounts of molten salt AlkX, (with the anion X^- acting both as ligand and solvent), the formed MX_m^{n-} complex may be characterized by two of the above-mentioned three attributes of the coordination compound (Table 2), namely by the coordination phenomenon, i.e. the unique arrangement of ligands around the central ion, and by a certain complex composition showing itself as the discrepancy between the formal degree of complex-forming metal oxidation and the coordination number. However, it is rather difficult to apply the third concept of heterolytical-type dissociation necessary for the complex to these systems, since the dissociation constants, (stability, instability), can be measured only in the case of the systems with an unambiguously-established stoichiometry of the complex-formation reaction and with a distinctive separation of the inner coordination sphere from the outer one, which is far from being evident for the case under review. (This specific feature applies to all the solvato-systems with coordinated solvents, including molten salts, and, therefore, the question concerning the possibility of introducing the concept of dissociation complex assumes fundamental importance). In such systems and concentration conditions self-oversolvation, (self-exchange of ligands), is encountered, and with a realization of the maximum coordination number of the central complex-forming ion in the condensated phase. It is self-evident that the reasonable equivalents of the third typical attribute of the complex compound, i.e heterolytic type dissociation constant, in such systems and concentration conditions should also be characterized with constancy, (in the case of unstoichiometry of the complex).

To support the validity of such a concept for melts the authors present the results of numerous studies of diluted, (less than 10% mol) solutions of $BeCl_2$, $TiCl_2$, $TiCl_3$, $ZrCl_3$, $HfCl_4$, VCl_3, $LaCl_3$, $CeCl_3$, $MoCl_3$ in molten chlorides IIIM which have shown a constancy in the coefficients of activity of these compounds within the studied concentration interval [10]. The same constancy is characteristic for the spectral parameters - coefficients of extinction and half-width of the electron absorption spectra (EAS) bands, and for

the values of magnetic susceptibility of the complex $NiCl_4^{2-}$ in melts as high as 10% mol MX_2 [8] (Figure 1).

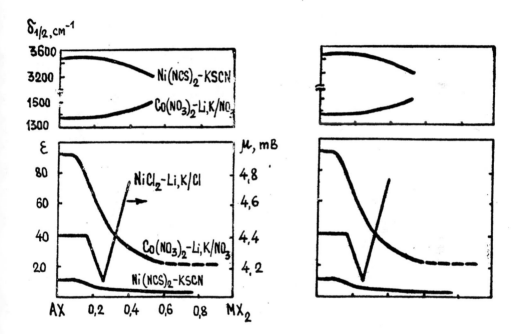

Figure 1. Coefficient of extinction ( ), half-width of EAS strips (1/2) and magnetic susceptibility (u) of various complex ions in molten salts of systems MX_2-AlkX as a function of composition.

The data on Raman scattering (RS) spectroscopy of the molten salt systems $XnCl_2$-KCl, $CdCl_2$-KCl, $MnCl_2$-KCl, $MgCl_2$-KCl are very convincing [11] and clearly show the constancy of the oscillation frequencies of M-Cl bonds up to the compound $X_{MX_2} \leq 0.33$, (i.e. in the diluted melts the only complex anion (in this case MX_4^{2-}) is also fixed and its parameters remain unchanged). All these, and other examples, indicate that in the region of the

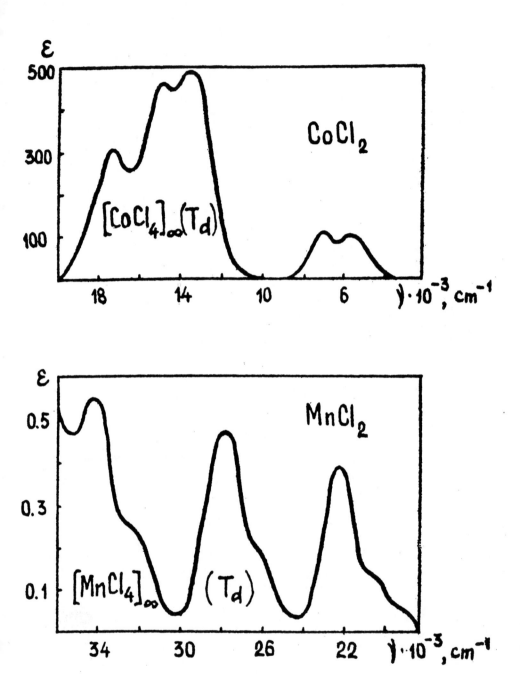

Figure 2. Electron absorption spectra (EAS) of molten $CoCl_2$ and $MnCl_2$ at 700°C.

diluted melts the unified discrete complex ion of maximum coordination saturation exists, for which it is basically possible to introduce the concept of heterolytical-type dissociation [ ]. The "lifetime of structural unit" concept introduced by D. Bokriss et al. for molten salts and determined as $1/K_i$, where K_i is the constant of complex dissociation rate [12] measured by galvanostatic and spectroscopic (RS) methods, may be taken as such an alternative concept for the solvate-complex systems.

Now let us discuss the second characteristic region of *concentrated solutions of ions M* in the system with a deficit of ligand X^-. For the systems of the MX-AlkX type it is logical to assume that in the MX-AlkX concentration region the gradual transfer from the structure of the above-mentioned complex ion MX_4^{2-} to the structure of individual molten salt MX will take place (taken as the extreme case). For retaining the coordination saturation of the central complex-forming ion, the logical idea of forming chain reticulate multinuclear, (in the extreme case - homonuclear), cluster structures of indefinite extension with the type bridge ions is introduced. This model is convincingly confirmed both by studies of EAS of individual molten salts $MnCl_2$, $CoCl_2$8 (Figure 2), (all central ions Mn^{2+} and Co^{2+} of the complex-forming metal are tetrahedronically surrounded by chlorines), and the neuronographical studies of the structures $ZnCl_2$ and $MgCl_2$ known from the literature [13].

The above-mentioned idea is also supported by the studies of magnetic susceptibility of the molten salt systems $MnCl_2$-AlkCl(Alk-Li, Na, K, Rb, Cs): if in the region of molten $MnCl_2$ salts the occurrence of the complex $MnCl_4^{2-}$ion is observed, then in the individual molten $MnCl_2$ the tetrahedronic indefinite-extension reticular structure will be formed with a high degree of certainty, with dimeric clusters Mn_2Cl_6 [11] being observed in the intermediate zone.

Thus, if even in these systems and concentration conditions these particles exhibit the features of a complex compound, (namely, the phenomenon of coordination and complex composition), they cannot be regarded as isolated complex ions of the definite composition. And, consequently, the compulsory parameter for the complex, namely the constant of instability of dissociation or lifetime or other similar but also constant parameter, does not apply to them. In this concentration region not a single parameter, magnetic susceptibility, coefficient of extinction, half-width of EAS bands, RS frequency, thermodynamic activity, is constant and the concept of thermodynamic or kinetic constants has no sense for such a set or for different or variable-composition multinuclear cluster groups.

On the basis of spectroscopic, magnetic, diffraction and thermodynamic studies of the molten salt systems with complex formation the authors proposed the complex-cluster model of their structure [8] (Figure 3). This model is based on the analysis of the equilibrium properties and stems from the ideas of dominant structural groups for each specific region of the compound.

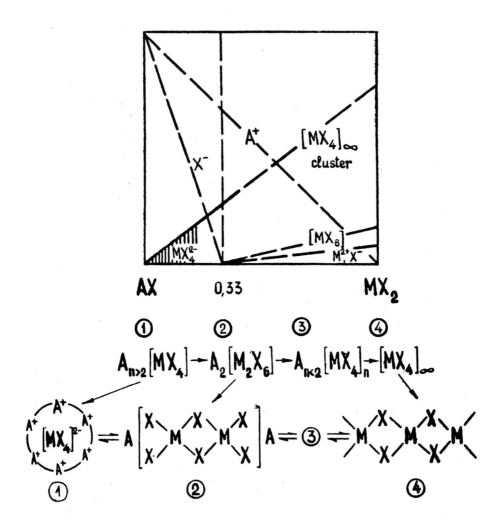

Figure 3. Complex-cluster structural model of molten-salt systems with complex formation: 1) region of discrete undistorted complex ions; 2) region of discrete distorted complex ions; 3) region of di- and hetero-nuclear complexes-clusters; 4) region of homonuclear clusters transforming into individual molten salt structure.

The region $0 < X_{MX} \leq 0.1$ is considered that which, along with non-coordinated anions X^- and cations Alk^+, contains the complex MX_m^{n-} anions of the maximum coordination saturation not distorted by the spherical field of the outer-sphere cations Alk+.

As the concentration of complexes increases, in the field $0.1 \leq X_{MX} \leq 0.33$, (in the case of complexes MX_m^{n-}), the conditions of their complex isolation from one another by the sufficiently spherical field of the outer-sphere cations are no longer satisfied and the already-distorted discrete complex anions are mainly observed.

The model of the structure of melts in the region $0.33 \leq X_{MX} < 1$ stems from the ideas of formation of structural cluster-type complexes which are predominant for these conditions and which are gradually becoming homonuclear with the same maximum coordination saturation of the MX_m cell.

The works of O'e et al. on IR and RS spectroscopy and of Blunder on molecular dynamics [13] of $AlCl_3$-AlkCl melts confirm the complex-cluster structural model and show that up to the ratio of 1:1 between components only the $AlCl_4^-$ complex takes place in this system, whereas at $X_{AlCl_3} > 0.5$ the clusters $Al_2Cl_7^-$, $Al_3Cl_{10}^-$ etc. become predominant particles.

Using RS spectroscopy and X-radiography methods Nakamura et al. [13] has confirmed that in the system $ZnCl_2$ - KCl up to $X_{ZnCl_2} = 0.33$ only an isolated tetrahedral anion $ZnCl_4^{2-}$ takes place, and at higher concentrations of $ZnCl_2$ - the cluster $Zn_2Cl_7^{3-}$ with linear-bridged chlorine which is regenerated into the polymer structure of individual zinc chloride with the same $Zn_2Cl_7^{3-}$ fragment, but with an angular arrangement of the bond. Thus, the complex-cluster structural model of molten salts with complex formation has found an unambiguous confirmation. However, the most important conclusion which can be made from this section is that along with the cluster formation zone the molten salt systems under discussion contain the region of discrete complex formation characterized by thermodynamic and kinetic constants.

2. DISCRETE COMPLEX ION IN MELTS AS A QUADRINUCLEAR HETERONUCLEAR SYSTEM

At first glance it may seem that having disclosed the conditions of existence of discrete complex ions and having outlined the boundaries of their domain the entire mechanism of the modern theoretical and experimental laws and relationships can be applied to them. However, the authors state that this is far from true.

It is sufficient to compare, for instance, the instability constants of complex ions in melts and aqueous solutions (Table 4) to make sure that their thermodynamic stability characteristics are quite different.

Though these data are not absolute, but carry only relative-character information on summary competitive processes of over-complexation, or so-called primary over-solvation: in melts - relative to nitrate (chloride) complexes and in water - relative to aquacomplexes, the analysis of Table 4 allows the following conclusions to be made. In most cases the strength of bromide, iodide, cyanide and ammonia complexes in aqueous solutions is higher than the strength of the same complexes in molten nitrates. Hence, it becomes clear that these aquacomplexes are not as strong as their nitrate complexes and the absence of the latter in aqueous solutions can be caused by the manifestation of the mass action law.

Table 4. Some constants of instability of complex ions in nitrate and chloride* melts and aqueous solutions [8, 14]

Complex ion	In Melt	In Water	Complex ion	In Melt	In Water
*CLF_2^+	$4\text{-}7.75 \cdot 10^{-3}$	$6.3 \cdot 10^{-4}$	$AgBr_2^-$	$3.2\text{-}7 \cdot 10^{-3}$	$7.8 \cdot 10^{-8}$
*LaF_2^+	$2.7\text{-}4.2 \cdot 10^{-3}$	$1.7 \cdot 10^{-3}$		$1.7 \cdot 10^{-5}$	$1.3 \cdot 10^{-9}$
	$3.23 \cdot 10^{-9}$	$1.40 \cdot 10^{-12}$	$AgBr_3^{2-}$	$1 \cdot 10^{-2}$	$2 \cdot 10^{-4}$
*$UO_2F_4^{2-}$	$2.1 \cdot 10^{-3}\text{-}6 \cdot 10^{-4}$	$1.7 \cdot 10^{-5}$	$CdBr_4^{2-}$	$2.8\text{-}4 \cdot 10^{-2}$	$1.2 \cdot 10^{-2}$
$AgCl_2^-$	$1.6\text{-}4 \cdot 10^{-3}$	$2.3 \cdot 10^{-2}$	$PbBr_2$	0.04	4
$PbCl_3$	$1.4\text{-}5.5 \cdot 10^{-2}$	$1.4 \cdot 10^{-2}$		$4.3 \cdot 10^{-8}$	$1.4 \cdot 10^{-14}$
$ZnCl^+$	$2.8\text{-}7.3 \cdot 10^{-4}$	0.19	$ZnBr^+$	$8 \cdot 10^{-9}$	$8\text{-}3 \cdot 10^{-7}$
	$4\text{-}6 \cdot 10^{-5}$	$3.4 \cdot 10^{-3}$	AgI_3^{2-}	$1 \cdot 10^{-3}\text{-}3\text{-}6 \cdot 10^{-4}$	$6.3 \cdot 10^{-4}$
$CdCl_3$	$1.5 \cdot 10^{-6}$	$9.3 \cdot 10^{-3}$	CdI_4^{2-}	$2.17 \cdot 10^{-7}$	$9.3 \cdot 10^{-8}$
$CdCl_4^{2-}$	0.37	$0.112\text{-}0.42$		$1.34 \cdot 10^{-11}$	$8 \cdot 10^{-22}$
	0.086	0.59	$Ag(NH_3)^+$		
$CuSO_4^-$			$Ag(NH_3)_2^+$		
$AgSO_4^-$			*$Ag(CN)_2^-$		

This conclusion would be flawless if the data presented in Table 4 were reduced to the unified value of the ionic force of the solution. But it is here that the chief cause of their discrepancy lies: the constants of instability of complex ions in aqueous solutions have been obtained at ionic force values of 0.5 to 5 (maximum) whereas in melts, at ionic force values of about 20 and upwards. But an increase in the value of the ionic force of the solution even by a unity (unities) causes a value of the complex stability constant to decrease by an order (orders) of magnitude. It is this information on which the following second conclusion is based: the chief cause of such a discrepancy between the complex ion instability constants in these media consists in the competitive action of the "bare" base-solution ions of the melt as compared with a milder action of solvated base-solution ions in aqueous and anhydrated solutions.

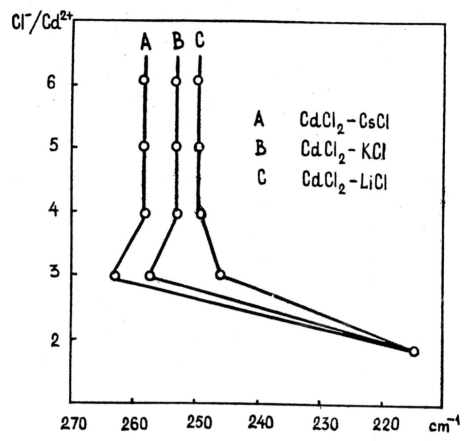

Figure 4. Relationship between oscillation frequency of bond Cd-Cl and ratio Cl/Cd_2 in a number of molten systems.

For the same reason, this summary of the macroproperty of many ions, should take the specific action of each type of the background cation into account. As an illustrative example, the authors compared the values of RS frequencies $v_1(A_1')$ of oscillation of the N-O bond in the nitrate ion in crystals, melts and aqueous solutions, (depending on the degree of dilution), of AlkNO_3 [15]. Irrespective of the salt, the hydrated NO_3^- in water is characterized for all the solutions by the oscillation frequency of bond N-O (A_1') equal to 1048 cm^{-1}, whereas in melts these frequencies differ and amount to 1067 cm^{-1} for LiNO_3, 1053 cm^{-1} for NaNO_3, 1048 cm^{-1} for KNO_3, 1045 cm^{-1} for RbNO_3 and 1043 cm^{-1} for CsNO_3.

A similar dependence takes place in the analysis of RS oscillation (A_1) of the Cd^{2+}-Cl^- bond in the $CdCl_4^-$ complex [16]. At high ratios of Cl^-/Cd^{2+}

in all cases, as was mentioned previously, the complex $CdCl_4^2$ ion maintains its individuality and stoichemistry and retains its characteristics unchangeable in each individual system of melts. At the same time, proceeding from the CdCL$_2$-CsCl system to the CdCl$_2$-KCl and CdCl$_2$-LiCl systems make it possible to note pronounced changes in frequency response of the Cd^{2+}-Cl^- bond (Figure 4).

Thus, the above-mentioned experimental data show that direct interaction of complex anions and outer-sphere cations as compared with aqueous and anhydrous solutions.

Now we shall try to demonstrate that such an interaction is governed by changes in the degree of covalence of the bond metal-ligand under the effect o the outer-sphere cations of the melt. For this purpose, the authors discuss

the EAS of the complex $CoCl_4^{2-}$ in different media: in the solution HCL-H$_2$O, in the melts AlkNO$_3$-AlkCl and in the melts AlkCl (Alk-Li, Na, K, Rb, Cs) (Figure 5).

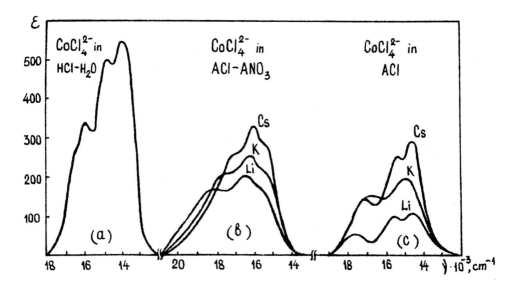

Figure 5. Electron absorption spectra of complex $CoCl^{2-}_4$ in HCl-H$_2$O solution (a) in melts Alk-Cl-AlkNO$_3$ (b); in melts AlkX(c) (where Alk-Li..., K..., Cs)

If Ballhousen's and Liehr's theory is applied, [17] establishing the relationships between the transition oscillator force (f) and the mixture of the 3-d orbitals of metals with orbitals of ligands to the intensity of EAS bands of the complex ions in melts, then the well-known equation:

$$f = \frac{1-\alpha^2}{1-2S\alpha(1-^2)^{0.5}} \cdot 16.32 \cdot 10^{11} A^2 \cdot \nu$$

(S is the overlapping integral, A is the interligand interaction integral; a is the ionicity), clearly shows the dependence of the oscillator force on the degree of covalence of the bond metal-ligand as a function of the outer-sphere cation. This dependence is maximum for the $CoCl_4^{2-}$ complex in the aqueous solution of HCl where ions, (both complex and base solution), are fully hydrated; lies somewhere between extremes in the $AlkNO_3$-AlkCl melts because of partial solvation of the outer-sphere cations by a large number of the melt-solvent NO_3^- ions, and is minimum in the AlkCl melts because of direct interaction of the outer-sphere cations with the $CoCl_4^{2-}$ complex.

It is evident that the intensity of the interaction, i.e. the covalence of the bonding between the $CoCl_4^{2-}$ complex and the outer-sphere cations A^+ grows in the series

$$Cs^+ < Rb^+ < K^+ < Na^+ < Li^+$$

Though in the systems of molten salts with complex formation the complex-forming region is clearly defined, it would be more correct to consider the complex anions *as quasiheteronuclear, quasi-binuclear particles of the Alkn[MX$_m$] type. Their parameters in solutions differ from the parameters of identical particles in aqueous and anhydrous solutions* due to the direct interaction between the complexes and the outer-sphere cations competing "in favor of ligands," which makes it impossible to merely apply them to the theoretical and experimental laws and relationships valid for conventional complexes. Hence, the need arises for developing them for the quantum-chemistry theories, such as theories of structure, theories of electron, oscillatory, Raman transitions and others.

3. QUANTUM-CHEMISTRY AND EXPERIMENTAL QUESTIONS OF STRUCTURE AND RADIATIVE AND NONRADIATIVE TRANSITIONS OF COORDINATION COMPOUNDS IN MOLTEN SALTS

The aim of this section of the paper is to describe the fundamental concepts of the quantum chemistry of coordination condensed systems now being developed to be applied to the complexes in molten salts. The authors will discuss such properties as quantum radiative transitions: electron, oscillatory, spin electron and nuclear, and nonradiative transitions depending on the composition and structure (symmetry) of such quasi-binuclear complex [9].

a) Electronic structure of coordination clusters of d-metals and electron transition spectra in molten salts. As was seen in Figure 5, the EAS, (frequencies, intensities), of complex ions in molten salts undergo changes depending on the outer-sphere cations. For the sake of greater clarity, the EAS parameters, (transition energies, classification of absorption bands, absorption band intensities - molar extinction coefficients), and the electronic structure parameters, (crystal field parameter - $10D_q$, nepheloxetic effect - b characterizing the degree of ionic bonding in the complex), for complex $MnCl_4^{2-}$, $CoCl_4^{2-}$ and $NiCl_4^{2-}$ ions in the binary MCl-AlkCl melts (Alk - Li, Na, K, Rb, Cs) was tabulated in Table 5. It is self-evident that the crystal field theory (CFT) parameter - $10D_q$ and the degree of ionicity b of the complex shall decrease in case of transfer from the outer-field cation Li^+ to Cs^+. Within the framework of the CFT, LCAO and LCMO the authors have developed the quantum-chemistry theory of the effect of the outer-sphere environment in the condensed systems on structure and spectral characteristics of complex.

Table 5. CFT parameter 10Dq (in LFT-D) and degree of bond ionicity b (nepheloxetic effect) of some complex ions in molten AlkCl as functions of outer-sphere cations Alk^+

Outer-sphere cation	10 Dq (D), cm^{-1}			$b=\dfrac{B}{B_0}$		
	$MnCl_4^{2-}$	$CoCl_4^{2-}$	$NiCl_4^{2-}$	$MnCl_4^{2-}$	$CoCl_4^{2-}$	$NiCl_4^{2-}$
Li+	5000	4100	–	0.74	0.70	–
Na+	4870	3900	4350	0.73	0.69	0.78
K+	4730	3700	4200	0.72	0.69	0.77
Pb+	4630	3500	4100	0.72	0.68	0.76
Cs+	4350	3250	4000	0.71	0.67	0.74

For instance, the problem of the effect of outer-sphere cations on the electron terms of complexes of d-metals is solved within the framework of the CFT by superimposition of the corresponding fields of the subsequent coordination fields [9]. As a result, the 10 Dq parameters may be written as follows:

D oct, tetr (cub)=Doct ± D' tetr (cub)

D tetr (cub), oct = Dtetr (cub) ± D^1 oct

D oct, oct = Doct ± D^1 oct

D tetr, tetr = Dtetr ± D^1 tetr'

where the subscripts "oct, tetr (cub)" denote octahedral, tetrahedral and cubic fields. The sequence of their writing corresponds to the sequence in which the inner and outer fields are superimposed on the central ion of the metal. The signs "+" and "-" denote positions of anions and cations, respectively.

Thus, splitting the terms of the d-metals in complexes conforms to the following rules: the combination of the tetrahedral and cubic fields of the first coordination sphere and octahedral field of outer-field cations (or anions), as well as the combination of the p octahedral field of the first coordination sphere of cations (or anions) results in an increase (or decrease) in the value of the splitting parameter 10Dq (D), whereas the combination of the fields of the same symmetry, (tetrahedron-tetrahedron, octahedron-octahedron), decreases this parameter in the case of the outer-sphere cations and increases it in the case of anions. In the series of the outer-sphere ions Li^+, Na^+, K^+, Rb^+, Cs^+ or F^-, Cl^-, Br^-, J^- these effects progressively weaken with the increase in the distance to the outer-sphere cation (anion), which is observed in the experiment (Figure 5, Table 5).

Because of the limitations of the CFT now allowing for the overlapping of the orbitals of the central ion of metal and ligands, (and the outer sphere cations), this theory fails to be used for evaluating the change of the covalence of the metal-ligand bond under the effect of the outer-sphere cation and other characteristics.

The most generalized approach for taking all the parameters of electronic structure of complex ions in molten salts into account is the molecular orbitals (MO) method: within the framework of the LCAO for evaluating the inner complex anion and within the framework of the LCMO for evaluating the entire cluster with the outer-sphere cations [9].

This MO cluster (y_i) shall be expressed in the following way through the MO of the complex (j_i) and group orbitals of the outer-sphere ions (F_i) with appropriate coefficients: $y_i = a_1 j_1 + a_2 F_2$.

Here, j_i for the complex of the d-metal may be e- or t_2-, and F_i - group atomic orbitals, for instance of the S-type for AlkX. In the series of the outer-sphere Li^+, Na^+, K^+, Rb^+, Cs^+ cations the coefficient a_1 tends to 1, and the coefficient a_2 related to a_1 through the expression $a_1 = -a_2 S + \sqrt{1 - a_2^2(1 - S^2)}$ and reflecting the degree of covalence tends to zero, which reflects a more pronounced covalence of the bond in the inner-sphere complex and weakening of the complex anion-outer-sphere cations bond and is convincingly supported by the data presented in Table 5.

The covalence parameter (1-b) increases in this series for the $MnCl_4^{2-}$, $CoCl_4^{2-}$, $NiCl_4^{2-}$ complexes, whereas the parameter (10Dq) decreases.

Within the framework of the LCMO method, the general expression can be written as:

$$\sum \Delta = \Delta_1 + \Delta_E - \Delta_{T_2} + \delta_{T_2} - \delta_E$$

where: D_1 is the splitting parameter in the first coordination sphere:

D_E, D_{T_2} is the stabilization of electron energy on the appropriate complex orbital in external field;

d_E, d_{T_2} are bonding effects.

Thus, all the EAS parameters and the parameters of structure of complex quasi-dinuclear ions in melts find the unambiguous quantum-chemistry explanation and can be well predicted within the framework of the cluster approach.

Table 6. Some Raman and IR oscillation frequencies of complex particles as functions of outer-sphere cations

Outer-sphere cations	Raman and IR frequencies; cm^{-1}									
	NO_3^-	NO_2^-	ClO_4^-	CO_3^{2-}	SO_4^{2-}	VO_2Cl_2	$CdCl_4^{2-}$	$ZnCl_4^{2-}$	$AlCl_4^-$	$InCl_4^-$
Li^+	1064-1050	1353	960	1072-1064	988	–	250	320-290	488	315
Na^+	1053-1048	1348	946		965	928	256	–	490	–
K^+	1047-1045	1338	937	1050-1031	955	025	258	290-280	487	320
Rb^+	1046-1041	1323	936		943	–	–	283-280	–	–
Cs^+	1043-1039	1310	935	1040-1025	–	924	o	276	483	322
				1020	–					

b) Oscillation (infrared (IR) and Raman scattering (RS)) spectra of coordination compounds in melted salts. The experimental data on the oscillation spectrum of coordination compounds in melted salts [8, 15] indicate that the oscillation frequency of the metal-ligand bonds, (there is scant data on their constants in melts), depend not only on the type of the central atom, degree of its oxidation, other characteristics and the origin of the ligand, but also, with all other conditions being equal, on the surrounding outer-sphere (Table 6). As a rule, the oscillation frequencies of the metal, (or element), ligand bonds in the complex ions are greater in the case of the outer-sphere partner Li^+ and decrease in the series towards the outer-sphere Cs^+, whereas, with due regard for the redistribution of electron density between the inner-sphere complex and the vacant orbitals of outer-sphere cations, (maximum for Li^+ and minimum for Cs^+), one would expect quite the contrary dependence.

The theoretical analysis of force constants of bonds solely as a function of electron density r varying depending on the parameters of outer-sphere cations, fails to yield an unequivocal answer sometimes demonstrating proportional and sometimes inverse-proportional and extreme dependencies [9].

To give an explanation to these "wayward" experimental facts, other theoretical concepts, namely the effect of the outer-sphere cations on the kinetic constants of the oscillation frequencies of quasi-dinuclear cluster particles $Alkn[MX_m]$ and the effect of electronic excitations on the force constants of oscillation frequencies of the complex itself, have to be involved. In [9] the authors show that the oscillation frequencies (w_a) can be described by the following expression:

$$\omega_\alpha = \sqrt{K_{\alpha\alpha} / t_{\alpha\alpha}} \cong \sqrt{K_{\alpha\alpha} / M_\alpha}$$

which bears a resemblance to the formula of oscillation of a diatomic molecule. Here, K_{LOC} and t_{aa} are the force and kinematic constants, M is the group mass. The dependence of frequency on the kinematic constants t_{aa} is evident: as the group mass M_a of the atoms involved in the given oscillation increases, the kinematic constants grow in magnitude and, consequently, frequencies decrease.

The other problem is to clear out the effect of variation of the composition of the coordination cluster with participation of outer-sphere cations on variation of the force constants K_{aa} in the symmetry coordinates. Expanding the electron Hamiltonian $\hat{H}(r,q)$ into a series to an accuracy of the second order of the perturbation theory yields the following expression for the force constant of the metal-ligand bond:

$$K_{\alpha\alpha} = \left(\frac{\partial^2 V_{nm}}{\partial Q_\alpha^2}\right)_0 + \left[\left(\frac{\partial^2 V_{EA}}{\partial Q_\alpha^2}\right)\right]_{00} - 2\sum_m^1 \left[\left(\frac{\partial V_{EA}}{\partial Q_\alpha}\right)_0\right]_{0m}^2 /(E_m^0 - E_0^0)$$

depending on the energy of electronic excitations ($E_m^0 - E_0^0$) discussed previously. With all other conditions being equal, decrease in the $\Delta \cong (E_m^0 - E_0^0)$ value for the complexes in the outer-sphere cation series $Li^+>Na^+>K^+>Rb^+>Cs^+$ should cause the value of K_{aa}, (just as the frequencies), to decrease due to an increase in the value of the third summand in the expression for K_{aa}. The scanty data available on the force constants of bonds of the complex ions $MoO_4^{2-}, WO_4^{2-}, CrO_4^{2-}, PO_4^{3-}, SO_4^{2-}, CO_3^{2-}$ in the molten NaCl, KCl, RbCl, CsCl at 885°C fully confirm this theoretical analysis. For instance,

The same is evident in the data presented in Table 6 showing the bond oscillation frequencies in complex ions as functions of the background cations in the melt.

The problem of disclosing the relationship between the intensity of the oscillation spectra of complex ions in molten salts and the outer-sphere cations is also of interest. However, the experimental data on this problem, (the gathering of which has started only recently), are obtained mainly in the form of relative intensities of IR-bands and Raman lines (table 7) not supported by rigorous theoretical calculations.

The relationships predictive in character may become valuable, such, for instance, as the proposed formula for intensity, (absorption coefficient of the IR-bands of the cluster in the condensed phase) [9]. Since the absorption coefficient of the IR-bands is a function of the square of the derivative of the dipole moment u with respect to $Q_a : æ \cdot (\partial mf / \partial Q_a)^2$ the discriminant function of the cluster by assuming the outer-level cations as the LCMO of the inner complex and group orbitals of outer-sphere cations can be constructed.

Table 7. Relative intensity of some IR- and Raman frequencies of complex particles in melt as a function of outer-sphere cations

Outer-sphere cations	Relative intensity of IR			Raman frequencies		
	in IR-spectra			in Raman spectra		
	NO_3^-	SO_4^{2-}	MoO_4^{2-}	ClO_3^-	NO_3^-	$AlCl_4^-$
Li^+	0.36	66	–	104.6	64	0.85
Na^+	0.32	66	73	107.8	63	–
K^+	0.33	71	78	88.9	73	0.59
Rb^+	–	–	–	–	–	–
Cs^+	–	–	–	–	–	0.60

As a result:

$$\left(\frac{\partial \bar{\mu}}{\partial Q_\alpha}\right)_0 = \left(\frac{\partial \bar{\mu}_0}{\partial Q_\alpha}\right) + \sum_{ikr} n_{ik}\left(a_{ik}^2 - 1 + a_{ik}a_{ikr}s_{ikr}\right)\left(\frac{\partial < \varphi_{ik} |\bar{r}| \varphi_{ik}}{\partial Q_\alpha}\right)_0$$

Since in the case of maximum ionic interaction of the inner complex with the outer-sphere cation Cs^+ a_1 tends to 1, and the product $(a_{ik} \bullet a_{ikr})$ tends to 0, and in the case of the maximum covalent bond between the inner complex and the outer-sphere cation Li^+ $a_{ik}^2 = (a_{ik} \bullet a_{ikr}) \rightarrow 1/\sqrt{2}$, the absorption coefficient æ shall decrease with an increasing degree of covalence of the complex-outer-sphere cation bond in the $Cs^+ \ldots Li^+$ series which in some cases (Table 7) is experimentally supported.

c) EAS spectra of coordination compounds in molten salts 15. From the equation for the g-factor [9]

$$g_{\alpha\beta} = g_{se}\left(\delta_{\alpha\beta} - \lambda \sum_{m \neq 0} \frac{<\psi_c | l_\alpha | \psi_m><\psi_m | l_{e\beta} | \psi_0>}{E_m - E_0}\right)$$

where (E_m-E_0) can approximately be substituted by the difference (e_m-e_0) between the unielectronic levels corresponding to the molecular orbitals y_0 and y_m in the products in the base and excited states, the dependence of the g-factor on the degree of delocalization of electrons, (degree of bond covalence), and on the electron excitation energy (e_m-e_0) can be seen.

Hence, the effect of the outer-sphere $Li^+...Cs^+$ cations on the g-factor of metal clusters can be determined through determining the effect of delocalization of the inner complex electrons on these cations. For this purpose, it is necessary to replace to MO_s of the cluster with the MO_s of the complex (j_i) and group orbitals of these cations (F_i) the effects of which should be taken with the sign (+) owing to the binding character of the MO_s of the entire cluster. Since the covalence of this bond, (between the complex and the outer-sphere cation), decreases in the $Li^+... Cs^+$ series, it should result in increasing values of g-factors for the clusters of d-metals with $d^{m<5}$, i.e. the metals with the d-shell were filled by less than half $(l>0)$ and decreasing values of g-factors for the clusters of d-metals with $d^{m>5}$ $(l>0)$. But this would be the case, if (e_m-e_0) were constant, i.e. $(e_m-e_0)=const$.

In reality, however, the value of $(e_m-e_0) \cdot 10$ $Dq \neq const$ and for clusters of d-metals also decreases in the series of the outer-sphere cations $Li^+...Cs^+$ (or $Mg^{2+}...Ba^{2+}$). The presence of this value in the denominator of the second negative member of the equation results in this series in a decrease of the g-factor for d^m 5-metals (0) and an increase of the g-factor for d^m 5-metals (0). The experimental values of the g-factors for clusters of the 3-d metals in crystals confirm this relationship [ ].

Crystal	MgO	CaO	SzO	Crystal	LiCl	NaCl	KCl	CsCl
V^{2+}	1.98	1.968	1.959		2.178	2.16	2.18	2.217
Cr^{3+}	1.978	1.973	–	Cu^{2+}	2.039	2.07	–	2.013
Ni^{2+}	2.215	2.327	–		–	2.06	–	–

Regretfully, there are available data on the values of g-factors for clusters in molten salts, and the effect of the outer-sphere cations on the parameters of the EAS spectra may only be confirmed by the results of measurements of the width of lines (H) of the EAS spectra of manganese (n) in octahedral clusters of the following molten fluorides:

Melt	MnF$_2$	KMnF$_3$	RbMnF$_3$	CsMnF$_3$
DH$_{Tmelt}$G-s	205	180	150	110
dDH/dT, mGs/K		-180 to -50	-150 to -70	-110

Narrowing of the line width signal is caused by the weaker dipole-dipole interaction of Alk$_n$[MnF$_6$] clusters in melts when proceeding from the system with K$^+$ cations located on the second coordination sphere to the systems with large Cs$^+$ and Rb$^+$ stronger ionic-bond with the complex.

The similar indirect information is carried by the measurements of magnetic susceptibility of the molten salt systems MnCl$_2$-AlkCl (Alk-Li, Na, K, Rb, Cs).

If the Poling bond-type magnetic criterion is used, then the pronounced decrease in paramagnetism of the MnCl$_4$$^{2-}$ complex is observed in the cases of the molten systems with outer-sphere cations Li$^+$, Na$^+$ and K$^+$ rather than Rb$^+$, Cs$^+$ which is accounted for namely by more covalent bonds complex-outer-sphere cation in the former case than in the latter.

d) NMR spectra of ion associates in molten salts 15

Whereas in the diluted aqueous solutions no change in the chemical shift of the NMR spectrum of atom nuclei with J≠0 is observed as a result of the exchange of the background cations of the medium, this charge is evidently observed in the molten salt systems (Figure 6). This change is primarily accounted for by the change in the bond covalence, for instance [205]Tl-X(X-Cl, Br, J) under the effect of cations of the melt Alk$^+$(Li$^+$...Cs$^+$) the covalence of the X--Alk$^+$ bond of which in its turn changes inversely to the first relationship (Figure 6a). It is the change of the covalence component of the bond, rather than polarization, that is evidenced by the order of chemical shifts [205]Tl:Cl-<Br-<J$^-$.

The quantum-chemistry interpretation of the effect of the background ions of the melt on the chemical shift, i.e. magnetic shielding of nuclei in these systems is more complex because of the need for taking their effect into account both on the dimagnetic (s^d) and paramagnetic (s^P) components of the total chemical shift. In general, it is also necessary to take the delocalization of electron density into account from the upper MOs of the covalent-bonded associate groups j$_i$ to the group orbitals F$_i$ of the competing background partner-cations Alk$^+$ of the melt composed of the vacant outer orbitals, which shall result in the MOs of the cluster of the type: $\psi_i = a_{ik}\varphi_{ik} + \sum_k a_{ik},\Phi_{ik}$. And the total chemical shift shall be expressed by the general formula [9]:

$$\sigma = \sigma^d + \sigma^P = \frac{1}{2}\int \psi_i \frac{1}{2}\int_j^3 \psi_i du - \frac{1}{E_m} - E_0 \int \psi_i \sum_{kk'} \frac{L_k L_k'}{r_j^3} \psi_i du$$

From the afore-mentioned it follows that both the diamagnetic and paramagnetic (in module) shielding components increase with a decreasing degree of covalence of the associate-competing anion bond observed in the series Li^+, Na^+, K^+, Rb^+, Cs^+ (these components approach the limit values in the case of purely ionic character of interaction).

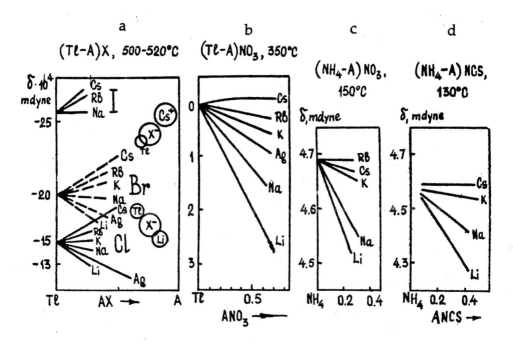

Figure 6. Chemical shifts of NMR of ^{205}Te, ^1H in molten salt systems as function of background cations: a) Tl-Alk)X, where X-Cl, Br, I; b) (TE-Alk)NO₃; (c) (NH₄-Alk)NO₃; d) (NH₄-Alk)NCs, where Alk-Li, Na, K, Rb, Cs.

In the $Li^+...Cs^+$ series, the paramagnetic shielding component should increase not only because of the decrease in the degree of covalence of the associate-competing cation bond, but also as a result of the decrease in the electron excitation energies (Em-E0) in the same series.

Thus, theoretically, the shielding $s=s^d+s^P$ in the series of the background cations of the melt from Li^+ to Cs^+ may increase, assume extreme values and even decrease depending on the ratio between the rates of the increase of the diamagnetic and paramagnetic (in module) shielding components (Figure 7a).

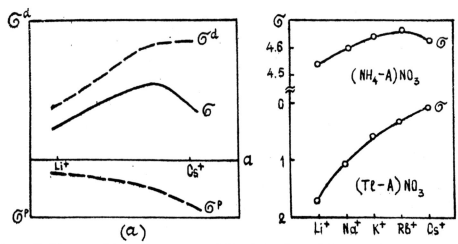

Figure 7. Theoretical (a) and experimental (b) relationships of chemical shifts of NMR as functions of background cations of melt $(NH_4-Alk)NO_3$ and $(Tl-Alk)NO_3$, where Alk-Li, Na, K, Rb, Cs.

The analysis of the experimental data [15,18] presented in Figure 7b unequivocally confirms these theoretical relationships.

e) Non-radiative transition - electron transfer between particles in molten salts

Unlike the previous discussion dealing with equilibrium, stationary parameters: radiative electron, oscillatory, spin-electron and nuclear transitions on one center, this section is devoted to a brief consideration of the kinetic problem - non-radiative transfer of an electron between two centers-clusters in molten salts. The effect of the background ions of the medium and the velocity of the outer-sphere electron transfer between complexes in aqueous and unhydrous media has been observed many times and may be illustrated for the set of particles similar to that under consideration in Table 8 which shows an increasing velocity of electron transfer between complex particles in the aqueous solution in the series of the background electrolite cations from Li+ to Cs+ [19].

Since an increase in the velocity of electron transfer in such solutions is thought to be caused by weaker electrostatic repulsion of the like complex ions due to the outer-sphere bridge cation Alk+, to explain the observe relationships one must assume that the outer-sphere Alk+ cations are arranged between the complex anions *jointly with* their hydrate, (in the general case, solvation), shells acting as such bridges [19]. This is why it is Cs+ with the minimum hydrate shell, rather than Li+, Na+ (and even K+ and Rb+), that ensures shorter electron transfer distances.

Table 8. Effect of outer-sphere cations on rates of reactions of electron transfer between particles in aqueous solution [19] and melts [20]

Background outer-sphere cation (10^{-1} to 1M)	$K(10^2$ cm0s^{-1}) in water			Background outer-sphere cation (10^{-1} to 1-2M)	$K(10^{-3}$ cm0s^{-1}) in melt	
	$MnO_4^{2-/-}$	$Fe(CN)_6^{4-/3-}$			$WO_4^{2-/4-}$	$NO_3^{-2/2-}$
	0.1M		1M			(170°C)
Li^+	7	–	7	Mg^{2+}	0.83--.42.7	2.8
Na^+	7.2	1.4	17			
K^+	8	2	2.8-24			
Rb^+	–	3	–	Ca^{2+}	0.23--1.27	0.36
Cs^+	24	5	23			

These electron transfer velocities are well correlated with the thermodynamical data: the stability of alkali metal cation associates - Alk+ with anion complexes $Fe(CN)_6^{3-/4-}$ increases in the $Li^+...Cs^+$ series at m=0.1 which confirms retaining of their hydrate shells [19].

However, the facts which are true for unconcentrated solutions (~0.1 to 1M) fail to be true in the high ionic-strength solutions (~3 to 4M): and both the afore-mentioned series of stability of contact pairs-associates of complex ions with outer-sphere cations and the series of their electrical potentials become transformed and, for instance, for $Fe(CN)_6^{3-/4-}$, take the form $K^+<Na^+<Li^+$ [19].

The same relationships, both thermodynamic and kinetic, and, primarily the same velocities of electron transfer between clusters are expected in the molten salts, due to a more pronounced direct contact of complex and other anions with the outer-sphere cations which has already been seen. Because of the data on the velocities of homogeneous electron transfer between complex particle in molten salts as a function of the outer-sphere cation was not available the authors will present only the available results (Table 8) for the velocities of the heterogeneous electron transfer in melts, which is quite legitimately because of a direct correlation between them [19].

As it is seen, the decrease in the velocity of the heterogeneous electron transfer for WO_4^{2-}, NO_3^- anions in molten salts, when proceeding from the background cation Mg^{2+} to Ca^{2+}, confirms the *a priory* assumption about the existence of such a relationship and is in agreement in this case with an increase in the distance between the particles in the melts. The same conclusions can be drawn from the data on the rates of more complex oxidation-reduction reactions in molten salts including the homogeneous electron transfer (Table 9): proceeding from the solvent with a large radius of cations to the solvents with a smaller radii of cations, all the investigated reactions are accelerated a few tens of times 21.

Since the probability of transfer (W), and consequently, the velocity of electron transfer (K) can be expressed as 9:

$$\bar{K} \sim W = \left(\frac{\pi}{h^2 kTE_p}\right)^{1/2} \left|V_{1_1 1_2}\right| \exp\left(-\frac{Ea}{kT}\right)$$

and are governed by the contributions of the following two parameters: the activation energy (Ea) and the transfer matrix element ($|V_{1_1 1_0}|$), let us analyze their relationships in this particular case: the complex anions in molten salts subjected to the effect of the outer-sphere background cations of the melt. The detailed theoretical analysis of electron transfer between a donor and an acceptor in the condensed medium is given in [9].

From the expression $E_a = \dfrac{(\varepsilon_{acc} - \varepsilon_{don} + E_r)^2}{4E_r}$ it is seen that the activation energy is a function of the electron energy levels on the acceptor (e_{acc}) and donor (e_{don}) and the medium reorganization energy (E_r). Under the conditions of localization of outer-sphere ions near the complex, (as a rule, with a smaller charge), and with all other conditions being equal, the electron energy level on the acceptor is stabilized in the case of outer-sphere cation (and the electron energy level on the donor is destabilized in the case of outer-sphere anions) [9]. In such a case, the activation energy of electron transfer should decrease when proceeding in the series from the outer-sphere cation Cs^+ to Li^+. In addition, in this series the medium (melt) reorganization energy increases as well (which is evident, for instance, from the data on the NMR of systems of melts TlX-AlkX and $TlNO_3$-$AlkNO_3$ (where X-Cl, Br, I; Alk-Li, Na, K, Rb, Cs) (Figure 6)) which results in a decrease in activation energy.

The afore-mentioned argument is convincingly confirmed by the data presented in Table 9 showing the activation energies of the oxidation - reduction reactions (1), (2), (3) which grow when proceeding from the outer-sphere cation Li^+ to K^+.

Apart from the activation energy, the electron transfer matrix element should also change in the field of the outer-sphere background ions of the melt. Since the transfer matrix element reduces the full potential of the matrix element (Coulomb and exchange) [9] acting on the electron being transferred in melts from the side of the complex anion, electron donor, complex anion - electron acceptor, outer-sphere cations and the remaining part of the medium, the outer-sphere cations creating additional accelerating potential in the intermediate region between anion particles increase the prexponential multiplier, i.e. also increase the electron transfer velocity. The electron transfer velocity, as seen from Tables 8 and 9, increases in proportion to the effective charge (z/r) of the outer-sphere cation amounting to 0.75 for K^+, 1.03 for Na^+, 1.28 to 1.67 for Li^+, 1.89 to 2.01 for Ca^{2+} and 2.56 to 3.08 for Mg^{2+}.

Table 9. Effect of outer-sphere cations on rates of oxidation-reduction reactions in melts 21

Solvent and background outer-sphere cation	K_{773K} (kg^{-1} M hr^{-1}) (kg^{-1} M hr^{-1}) (hr^{-1})			E(kJ◊M^{-1}) (1) (2) (3)		
	$1/4NO_2^- + ClO_4^-$	ClO_4^-	NO_3^-			
	$1/4NO_3^- + Cl^-$	$Cl^- + 2O_2$	$NO_2^- + 1/2 O_2$			
	(1)	(2)	(3)			
$\underline{Li}ClO_4$	139,06	–	–	129	–	–
$\underline{Li}, Na/NO_3$	–	12.57	–	–	206	–
$\underline{Na}NO_3$	53.19	3.45	$5.80 \cdot 10^{-3}$	206	200	185
$\underline{Na,K}/NO_3$	46.56	4.56	$2.15 \cdot 10^{-4}$	225	224	270
$\underline{K}NO_3$	42.38	1.72	$3.550 \cdot 10^{-5}$	249	209	275

4. SOME COMMON AND SPECIFIC PROBLEMS OF ION COORDINATION IN MOLTEN SALTS AND MOLECULES IN MOLECULAR HIGH-BOILING SOLVENTS

To make the transition to the next section of the paper, discussion of co-ordination compounds in anhydrous high-boiling (often, molecular) solvents, more logical the authors focus their attention on a number of specific features and features common with the reviewed complex compounds in molten salts [22].

It is self-evident that in both cases, i.e. in the condensed state, the coordination saturation of the bonds between complexes must take place and their maximally possible coordination numbers must be realized.

This easily interpretable principle of the "maximum coordination number" pre-supposes that in the aqueous and other like systems of solvents with high dielectric permeability a gradual transition should take place from an individual crystal, for instance $[MX_2]_\bullet$, to crystalline-hydrates (-solvates) of the type $MX_2 x_m$ (H_2O) or $x_n(S)$ and the like, and so on up to the individual aqua- (solvato-) complexes of the type M^{2+} (H_2O)$_6$ or M^{2+} $(S)_{p+2}$ in diluted solutions. It should be noted that the maximum coordination numbers are ensured either by the sum of the similar-type ligands, (including the bridging ones, which primarily manifests itself in crystals), or by the combination of different ligands. In such systems the authors deal with electrolytic dissociation of the dissolved substance and with complete dissociation with dilution of the complex to "resolvate" it for a more strong ligand (Table 10).

Table 10. Types of systems with coordinated solvent and basic processes occurring in them

Types of systems with co-ordinated solvent	Basic processes
Systems with high dielectric permeability of solvent (H_2O, CN, CHOOH, ...)	Electrolytic dissociation of solute and complete resolvation for a more "potent" ligand $$[MX_n]_\infty \cdot H_2O \rightarrow M(H_2O)_p^{n+} +$$ $$+nX^-(H_2O)m$$
Molten salts	Partial thermal dissociation and "completion of solvation" of solute $$\left. \begin{array}{l} (AX)_\infty \overset{\rightarrow}{\leftarrow} pA^+ + pX^- \\ MX_n \overset{\rightarrow}{\leftarrow} MX_{n-1}^+ + X^- \end{array} \right\} \begin{array}{l} MX_n, MX^+, M^n + p(X^-, A^+) = \\ = [MX_{n+p}]^{p^-} \cdot pA^+ \end{array}$$ "Solvation" of solute $MX_n + pSolv \mathbb{E} [MX_n(Solv)_p]$ Sometimes accompanied by thermal dissociation $AlCl_3 + SCl_4 \mathbb{E} [AlCl_3 \Diamond (SCl_4)] \mathbb{E} AlCl_4 + SCl_3^+$ and by electrolytical dissociation $$OsCl_4 + 2SCl_4 \rightarrow [OsCl_4 \cdot (SCl_4)_2] \overset{SoCl_2}{\underset{\longleftarrow}{\longrightarrow}} OsCl_6^{2-} + 2SCl_3^+$$

In the case of the solvents with low dielectric permeability one cannot expect the electrolytic dissociation of the solute and, as a rule, have to deal with the process of its "additional solvation" by the coordinated solvent to the coordinate number of the central complex-forming atom optimally maximum for this system and conditions. In turn, the complex-composition coordinated solvent may undergo further transformations caused by the primary coordination processes. Namely, because of a weakening of internal bonding the ligand may undergo further electrolytic and thermal dissociation, etc. (Table 10).

These sufficiently clear concepts, however, are not always correctly interpreted in the case of the molten salts systems. First, for instance, the presence on their phase diagrams even the congruently-melted compounds of the 1:1 composition, such as $CdCl_2$: KCl, $MgCl_2$: KCl, etc. are often unequivocally interpreted as the existence in the melt of the complex ions $CdCl_3^-$, $MgCl_3^-$ [23]. It should be noted however that the phase analysis yields only the ratio between 1:1 components (=2:2; =3:3, etc.), rather than their quantitative amounts, thus making such an approach illegitimate. For instance, despite the fact that the phase diagram of the $ZnCl_2$-KCl system shows a formation of the 2:1- and 1:2-composition compounds, in the melt, both the Raman spectra and the diffraction measurements confirm the presence within the entire concentration interval of the system of purely tetrahedronically coordinated ion $ZnCl_4^{2-}$ with end Cl^- - ligands in diluted solutions or bridging ligands in $Zn_2Cl_7^{3-}$, etc. in concentrated solutions [13]. This and the author's other data as well as data taken from the literature have enabled us to develop the complex-cluster model of structure of the molten salts with complex formation 8 which has established the way for obtaining the coordination numbers optimum for given systems and conditions.

Table 11. Static dielectric permittivity of some liquids, including melts and water, in various conditions

Liquid sub-stance		Liquid substance		Melt	
H_2O	80(~94)	H_2O(400°C, p≥1 kbar)	10-5	H_2O(~1000°C, p>150 kbar)	5
$HCONH_2$	90	$SOCl_2$	9.1	NaCl	~2 (6.12)
$(CH_2OH)_2$	41	S_2Cl_2	4.8-5.8	KCl	~2 (5.03)
CH_3CN	39	SCl_2	3-3.5	$GeBr_4$	2.95 (3.17)
$C_6H_5NO_2$	36	CS_2 CCl_4	2.63 2.23	$AlBr_3$	2.9

The second cause of misinterpretation of these systems lies in the *a priori*-postulated concepts of solely-dielectric dissociation of molten salts. However, as seen from Table 11, the molten salts are classified as the liquids with very low dielectric permittivity owing to which the partial dissociation of them (and in them) is of predominantly thermal character, whereas in the case of dilutes this partial dissociation takes place mainly in diluted solutions. Therefore, the complex-formation processes with coordinated solvents of such type are based on the thermal dissociation of part of the molecules of the solvent, (and less-molecules of the solute), and on the "desolvation" of the solute to the coordination number maximum for this system and these conditions either by individual ions, for instance F^-, Cl^-, Br^-, etc., or by more complex molecular formations of the type $AlCl_4^-$, $ZnCl_4^{2-}$, etc. depending on particular systems and conditions (Table 10). In the latter case one is dealing with the formation of heteronuclear complex groups.

Taking the concept of the *maximum* optimum coordination number for the selected conditions and systems as a basis and regarding the dielectric permittivity of solvents as an illustrative factor determining the character of various chemical processes the purposeful synthesis of various complexes can be predicted.

For instance, in the case of the generally-known synthesis in the systems with high-permittivity solvents, (H_2O, HF, H_2SO_4, etc.) it is necessary to take the electrolytic dissociation of solutes into account, as well as complete resolution of the complex-forming metal, (in case of dilution), towards the coordinated solvent more potent as a ligand, which results in a high solubility of the formed solvato-(aqua-) complexes (Table 10).

This same result, even though obtained in another way, (i.e. by "completion of solvation" of the solutes by thermally dissociated ions of the solvent), is reached in the systems of molten salts with complex formation (Table 10). However, in this case the greater freedom in manipulating the temperature factor allows the required substances to be often crystallized from melts in the form of the desired final product.

It is quite another matter, when complexes are formed in the system containing the solvent with low dielectric permittivity ($SOCl_2$, SCl_2, CS_2, CCl_4, etc.). First, the absence of or insufficient dissociation of solutes in them and second, restricted solvation capability of these solvents results in the formation of numerous weekly "solvated" products: adducts (Table 10) poorly soluble through the afore-mentioned reason in such, as a rule, anhydrous, molecular, (sometimes, high-boiling), solvents and, hence, readily crystallized in them.

In studying the systems of this type, normally distinguished with the formation of heteronuclear complexes that the following section of this paper is devoted to.

The author's already mentioned the formation of such complexes in the molten salt systems: the authors of [24] observed formation of heteronuclear chloride complexes of titanium, vanadium, chrome, manganese, iron, cobalt, nickel, copper (M^{2+}) with aluminum chloride: $M^{2+} + 2Al_2Cl_7^- \rightarrow M(Al_2Cl_7)_2$; in [8, 15] the authors found the heteronuclear chloride complexes of iron and tin, manganese and zinc, manganese and cadmium of type $M_n(CdCl_3)_2$, Fe (S_4Cl_3), etc.

It is the absence of the coordination solvents with high dielectric permittivity separating their solvated "overcoat" the dissociating individual metal complexes, on the one hand, and the high coordination saturation of Lewis acids (halides of aluminum, zinc, etc.) , on the other hand, that creates favorable conditions for obtaining these heteronuclear coordination compounds. For example, the synthesis of numerous heteronuclear bromide complexes of chromium, manganese, cobalts, nickel, palladium, zinc, cadmium and mercury (M) with aluminum bromide in toluene and cyclohexane [25]: MBrs(solid)+Al2Br6(liquid) MAl2Br8(liquid).

III. CHEMISTRY OF COORDINATION COMPOUNDS IN ANHYDROUS HIGH-BOILING SOLVENTS

In this section[*] the authors make an attempt to focus their attention on the following three basic problems of coordination chemistry in molecular (anhydrous) (including high-boiling) solvents:

- simultaneous exhibition of properties of a solvent, ligand and reagent;
- formation of heteronuclear, dinuclear and cluster complexes in them and desired synthesis of such complexes;
- some solvatometallurgical aspects of the problem.

Because of a large number of such systems and the limited space allowed for this paper, the authors will try to discuss the afore-mentioned problem taking as an example the traditional studies of complex formation of metals in chalcogen halide media attracting enormous interest as new reaction-capable "low-temperature" media that not only possess the characteristics of anhydrous solvent, but are capable of forming complexes and causing a chlorinating oxidation effect.

Since their potential application has been dealt with in the non-ferrous metal chemistry and non-ferrous metallurgy, the authors consider the problems of complex formation by noble metals [26-36].

1. PROBLEM OF SIMULTANEOUS EXHIBITION OF PROPERTIES OF SOLVENT-LIGAND-REAGENT

First, the authors will give the brief characteristic of chalcogen halides themselves [27]. Halogens and chalcogens form a number of compounds with a halogen-chalcogen (X-E) ratio of 1:1; 2:1 and 4:1. Chain-like polymer forms also known as C_nH_2, where n=2-8. Lengths of the bonds H-C are equal to 2.01; 2.24; 2.16; 2.36A for chlorides and bromides of sulphur and selenium, respectively, and are close to simple covalent bonds. In tetrahalide molecules one bond is considerably longer (up to 3A in the molecule $SeCl_4$). The solid tetrahalides of selenium and tellurium are essentially tetramers $(EX_4)_4$ with one extended bridging bond. Dihalides EX_2 have the angular structure (<100°) (symmetry C_{2V}) and monohalides are of bridging structure with a dihedral angle of 83 degrees (symmetry C_2). Such a geometry of chalcogen halides reflects the SP^3-hybrid state. These specific features of the structure of electrically-saturated covalent molecules of chalcogen halides strongly affect their chemical and physical properties. Low halides are essentially low-polar liquids within wide temperature limits (from -80°C to +138°C for

---

[*] This section has been prepared with the assistance of V.A. Grafov, V.I. Pekhnyo, Z.A. Fokina and others.

S_2Cl_2); tetrahalides of sulphur and selenium are distinguished by thermal instability and decompose with the evolvement of active halogen. Sulphur tetrachloride melts with decomposition at -31°C; $4SCl_4 \rightleftarrows 2SCl_2+S_2Cl_2+5Cl_2$; selenium tetrachloride sublimes and transforms into a gaseous phase with decomposition at 175°C; $SeCl_4 \rightleftarrows SeCl_2$ (80%) + Cl_2(20%); tellurium tetrachloride melts and boils at 224°C without decomposition; thermal stability of tetrabromides is still lower. Selenium and tellurium halides in the gaseous phase predominantly exist in the form of dihalides; sulphur dichloride exists in the condensed phase (t_{melt}=122°C, t_{boil}=59°C), but in the absence of stabilizers decomposes according to the reaction $2SCl_2 \rightleftarrows S_2Cl_2+2Cl$.

Having small enthalpies of formation, (58.7 and 49.4 kJ/mole for S_2Cl_2 and SCl_2, respectively), sulphur chlorides are essentially sources of active chlorine. Thus, in the S-Se-Te series the stability increases for the tetrahalides and decreases for mon-halide forms; tellurium monohalide and sulphur tetrabromide have not yet been obtained. The full set of thermodynamical characteristics is not available for chalcogen halides; however, low enthalpies of formation and the fact that these enthalpies are close in magnitude for various compounds, (E_2X_2, EX_2, EX_4), suggests the presence of the equilibrium state of these and other forms within wide ranges of temperatures and concentrations and, consequently, allows the chalcogen halides to be regarded as active halogenating components.

The complex-formation reactions are essentially untypical for the chemistry of low chalcogen halides; the characteristic features of chalcogen tetrahalides are amphoterism and complex formation following two types of reactions: (1) formation of anion forms $(EX_5)^-$, $(EX_6)^{2-}$ and (2) formation of cation forms $(EX_3)^+$. In the series of S-Se-Te halides, the acceptor properties grow: for sulphur no complex formed by reaction (1) is known, while complexes formed by reaction (2) are known for all tetrahalides with halides of non-transition P-metals-Lewis acids. In these compounds the chalcogen tetrahalide is coordinated through the agency of the halogen atom and this coordination may be accompanied by complete dissociation $EX_4 \rightleftarrows EX^+_3+X^-$ with formation of ?? and complex metallanion MX_m^{n-m}. There is not much data on the coordination of chalcogen dihalides, the only known structure is the planar square structure of compounds of tellurium with thiourea TeX_2 $(ThiO)_2$ which confirms that the coordination bond is formed with the participation of 5d-orbitals with the contribution of acceptor properties of tellurium.

Until recently, no complex compounds of chalcogen halides with platinum metals were known. Numerous attempts to obtain these compounds, (Groneveld, 1953; Romankevich, 1959; Bedly, 1974; Korshunov, 1980; Denika, 1980), have ended in failure. The only known example now is the gold complexes $AnSCl_7$ and $AnSeCl_7$ which were obtained by Linde in 1885 and whose structure and properties have not yet been studied.

Negative values of free energies for most reactions of chalcogen chlorides with noble metals (up to 500°C) suggest that in principle noble metals can form chlorides when reacting with sulphur chlorides.

The exception here is the reactions of chlorination of gold with sulphur monochloride:

An+0.5 S_2Cl_2 AnCl+S(DG=4.2 kJ/mole),
An+1.5 S_2Cl_2 AnCl3+3S (DG=10.5 kJ/mole)

where the equilibrium shifted to the left.

The analysis of the electronic structure of sulphur chlorides has shown that sulphur tetrachloride has the atoms of Cl with a high negative charge (-0.41 e) and that in the dichloride molecules the changes on the sulphur and chlorine atoms are distributed rather uniformly (+0.18e and -0.09e). At the same time, much contribution to the highest occupied molecular orbital in the dichloride molecules is made by the unshared electron pair (3p) of the sulphur atom; this orbital of the atom of sulphur is of mixed character. Taking these specific features of the electronic structure into account, chalcogen tetrahalides can be expected to exhibit the predominant coordination through the interaction of the atom of the halogen with the high-charge cations of platinum metals (Os^{IV}, Ru^{IV}, Pt^{IV}), i.e. "charge-controlled" interaction. For chalcogen dihalides the "orbital-controlled" interaction should be expected, i.e. the interaction of the atom of the chalcogen with platinum metals with the lowest degrees of oxidation (Pt^{II}, Pd^{II}, Ir^{III}, Rh^{I}) having relatively low located lowest unoccupied molecular orbital.

The analysis of reaction capacity has allowed us to select the necessary conditions for conducting complex-formation reactions and accomplish synthesis of the compounds of the new class - halogen - chalcogenide complexes of noble metals [26, 27]. Taking the physical and chemical properties of chalcogen halides into account the following recommendations for synthesis have been determined:

- selection of unhydrous solvent: use of the haloid-donor solvents, (thionyl chloride, sulphur dichloride, bromine), results in the coordination of the tetrahalide forms of chalcogens; use of the "indifferent" solvents, (methylene chloride, sulphur monochloride), ensures coordination of dihalide forms of chalcogens;
- an increase in the synthesis temperature (above 170°C), results in the coordination of dihalide forms of chalcogens which are more stable under these temperatures;
- creation of an "active" form of metal owing to the application of its original aqua- (hydroxo⁻) halide forms because of reactory inertia of the simple halides of noble metals having the cluster nature;
- formation of the ligand SeX_2 (not existing in the condensed phase in normal conditions) due to utilization of the equimolar relationships SeX_4+Se;

Table 12. New Halogenhalcogenide complexes of noble metal [27, 36]

Nos	Compound	t_{decomp} (beginning) °C	Elementary cell parameters				$\mu_{eff.} / V$	syngony
			a, Å	b, Å	c, Å	z		
1	PdS_2Cl_6	80	6.18	7.54	7.11	2	0	
2	$PdSe_2Cl_6$	115	9.09	6.61	8.21	2	0	
3	$PdTe_2Cl_6$	–	8.13	14.47	8.51		0	rhombic
4	$PdSe_2Br_6$	165	7.21	9.21	8.43	2	0	
5	PdS_2Cl_8	–	X-ray amorphous					
6	$PdSl_2Cl_{12}$	–	13.71	14.47	12.25		0	rhombic
7	$PdTe_2Cl_{12}$	–	9.76	7.24	11.11		0	rhombic
8	PtS_2Cl_8	105	6.16	9.12	5.16	1	0	rhombic
9	$PtSe_2Cl_8$	125	6.08	8.43	12.49	2	0	rhombic
10	$PtSe_2Cl_{12}$	120	10.61	10.61	7.83	2	0	tetragonal
11	$PtTe_2Cl_{12}$	245	7.41	7.41	9.85	1	0	tetragonal
12	$PtSe_2Br_{12}$	145	8.10	11.15	6.06	1	0	
13	$PtTe_2Br_{12}$	–	8.96	8.22	13.85	2	0	
14	OsS_2Cl_{12}	75	7.80	5.71	10.65	1	1.34	rhombic
15	$OsSe_2Cl_{12}$	110	10.52	10.52	7.68	2	1.40	tetragonal
16	$OsTe_2Cl_{12}$	250	8.88	8.88	9.84	2	1.45	tetragonal
17	RuS_2Cl_7	–	8.75	10.83	12.41		0	rhombic
18	$RuTe_2Cl_{12}$	175	13.73	8.16	8.79	2	0	rhombic
19	$RuSe_3Cl_9$	155	6.44	6.44	9.62	1	0	tetragonal
20	RhS_3Cl_9	–	X-ray amorphous				0	
21	$RhSe_2Cl_7$	–	7.88	10.77	12.97		0	rhombic
22	$RhTe_3Cl_{15}$	–	X-ray amorphous				0	
23	IrS_2Cl_7	80	12.11	–	7.96	2	0	
24	$IrSl_2Cl_7$	110	low-syngony				0	
25	$AuSCl_7$	115	10.66	–	7.76	4	0	
26	$AuSeCl_7$	120	low syngony				0	
27	$AuTeCl_7$	200	low syngony				0	
28	$AuSCl_5$	112	7.26	9.67	8.27		0	
29	$AuSeCl_5$	132	6.90	6.51	11.86		0	
30	$AuSeBr_7$	120	low syngony				0	
31	$AuTeBr_7$	–	low syngony				0	

-   conducting of the synthesis in sealed vials in order to prevent hydrolysis of chalcogen halides and formed complexes and to allow the synthesis to be performed at higher temperatures (up to 250°C), with the solvent vapor condensation at the cold end of the vial.

The normal practice is to conduct the synthesis of complexes at temperatures of 25 to 250°C, with the obtained compounds being recrystallized and dried first in the flow of chlorine (bromine) and then in a vacuum. The composition is established by the analysis of all the components: metal, halogen and chalcogen; density is determined in carbon tetrachloride; melting points (decomposition temperatures), are determined thermographically; crystallographic characteristics are established by X-ray phase studies (Table 12).

The reactions for obtaining complex halogen-chalcogenide compounds of noble metals can be classified as follows:

a) *Direct reaction of components.* With use of this method the following complexes have been obtained from salts of metals and ligands:

$$AnCl_3 + SeCl_4 \, (or \, TeCl_4) \xrightarrow{\text{Solv.}} [AuCl_3 \, (\ni Cl_4)],$$

$$OsCl_4 + 2SeCl_4 \, (or \, TeCl_4) \xrightarrow{\text{SOCl}_2} [OsCl_4 \, (\ni Cl_4)],$$

$$PtCl_4 + 2SCl_2 \longrightarrow [PtCl_4 \, (SCl_2)_s],$$

$$PtCl_4 + 2SeCl_4 \, (or \, TeCl_4) \xrightarrow{\text{SOCl}_2} [PtCl_4 \, (\ni Cl_4)_s],$$

$$PdCl_2 + 2SCl_2 \longrightarrow [PdCl_2 \, (SCl_2)_2],$$

$$AuBr_3 + SeBr_4 \, (or \, TeBr_4) \xrightarrow{\text{Br}_2} [AuBr_3 \, (\ni Br_4)],$$

$$PtBr_4 + 2SeBr_4 \, (or \, TeBr_4) \xrightarrow{\text{Br}_2} [PtBr_4 \, (\ni Br_4)_2].$$

Due to a number of limitations of this method arising from instability of ligands under normal conditions and the inability to find the complex that would equally solve both ligands and metal halides, most of the compounds have been obtained by the methods outlined in the following text.

b) Apparent reagents method based on the formation of the system of ligands or necessary salts of metals in the course of the synthesis. This method has been used for obtaining the ligands unstable in normal conditions:

$$AuCl_3 + SCl_2 \xrightarrow{\text{SOCl}_2} [AuCl_3 \, (SCl_4)],$$

$$PdCl_2 + SeCl_4 + Se \xrightarrow{\text{SOCl}_2} [PdCl_2 \, (SeCl_2)_2],$$

$$PtCl_4 + SeCl_4 + Se \xrightarrow{\text{SOCl}_2} [PtCl_4 \, (SeCl_2)_2],$$

$$PdBr_2 + Se_2Br_2 + Se \xrightarrow{\text{Br}_2} [PdBr_2 \, (SeBr_2)_2],$$

$$OsCl_4 + 2SCl_2 \xrightarrow{\text{SOCl}_2} [OsCl_6](SCl_3)_2,$$

for obtaining active intermediate form of salts and metals

$$IrCl_4 \bullet nH_2O + SCl_2 \longrightarrow [Ir_2Cl_6(SCl_2)_4],$$

$$IrCl_4 \bullet nH_2O + SeCl_2 \xrightarrow{SOCl_2} [Ir_2Cl_6(SeCl_2)_4],$$

$$RuCl_3(OH) + TeCl_4 \xrightarrow{SOCl_2} [RuCl_4(TeCl_4)_2],$$

$$OsO_4 + 2SCl_2 \longrightarrow [OsCl_6](SCl_3)_2,$$

$$OsO_4 + 2SeCl_4 \xrightarrow{SOCl_2} OsCl_4(\ni Cl_4)_2.$$

$$(\text{or } TeCl_4)$$

c) *The ligands exchange method* used for obtaining more stable selen- and tellurium-containing compounds from readily-available and easy-to-synthesize sulphur-containing analogs. This method of synthesis can readily be used for the metals forming series of complexes with all chalcogens in the similar haloxide forms and has been used for obtaining gold and osmium compounds:

$$[AuCl_3(SCl_4)] + SeCl_4 (\text{or } TeCl_4) \rightarrow [AuCl_3(\ni CL_4)] + SCl_2 + Cl_2$$

d) *Direct oxidative synthesis method.* This method has been used for obtaining sulphur-containing compounds of gold, palladium and platinum. The synthesis proceeds as follows:

$$M° + pL(= Solv. \equiv Ox) \rightarrow [M^{n+}L_m] + (p - m)L(\equiv Solv. \equiv Red.),$$

where the sulphur chlorides serve three functions concurrently: as solvent (Solv), as oxidizer (Ox) and as ligand (L). Conversion Red⇄Ox is easily (25°C) performed with the absorption of chlorine, which enables multiple use of sulphur chloride and precludes formation of side products. From the high-yield (100%) reaction medium the high-purity complexes are crystallized:

$$2\,Au + 12SCl_2 \rightarrow 2[AuCl_3(SCl_4)] + 5S_2Cl_2$$

$$Pd + 4SCl_2 \rightarrow [PdCl_2(SCl_2)_2] + S_2Cl_2$$

$$Pt + 6SCl_2 \rightarrow [PtCl_4(SCl_2)_2] + 2S_2Cl_2$$

$$2\,Au + 2Se + 7Br_2 \rightarrow 2[AuBr_3(SeBr_4)].$$

The difficulties in unambiguous determination of the structure of the new halogen-chalcogenide complexes of noble metals are aggravated by the likely existence of several stages of metal oxidation, a probable large variety of halides being coordinated, possible formation of dinuclear bridged complex forms, etc. Hence, despite the relatively simple bulk chemical composition of these halogen-chalcogenide complexes the authors are unable to determine their structure on the basis of analytical data. This problem was solved by using the spectroscopic research methods; infrared, Raman and nuclear quadropole resonance (NQR) spectroscopy on the atoms of chlorine and bromine; electron spectroscopy; X-ray spectroscopy (KB$_1$-spectra of chlorine and sulphur) and by measurements of magnetic susceptibility. In the research, use was made of such installation as IKS-14A with prism CsJ (200-400 cm^{-1}), Hitachi FIS-3 (60-400 cm^{-1}), Spex Ramalog (Ar-laser, 5145Å), DFS-24 (He-Ne-laser, 6328Å)-vibrational spectra; ISSh-1, ISP-1, ISSh-1, ISSh-2-13 (NQR spectra); PU-8800 (190-800 nm), Schimadzu RSF-7B (230-2500 nm) - electron diffuse reflection spectra. Magnetic susceptibility was measured on the installation following the Faraday-Sexmith method.

Spectroscopic characteristics have made it possible to establish the degree of oxidation of the central atom, its electron configuration, coordination number and geometry of coordinate site. The specific features of coordination of molecules of chalcogen halides - the form of coordinated halide, the donor atom in the ligand, the question relating to isomerism, (geometrical and linkage), were studied by IR$^-$, Raman- and NQR-spectroscopy methods.

2. FORMATION OF HETERONUCLEAR, DINUCLEAR AND OTHER COMPLEXES

The NQR spectra of the halogen atoms has turned out to be highly informative in studying the construction of halogen-chalcogenide complexes of noble metals. The NQRs under review were studied within the framework of the Towns-Daily model adjusted to the atoms of halogens with axial symmetry of the electrical field set up by the valence p-electrons.

The parameters discussed in interpreting the resultant NQR spectra of ^{35}Cl and ^{79}Br atoms (See Table 13) are the position of the frequency in the spectrum (conclusions on the character of the chemical bond), the character of the multiplet (difference between the geometrical isomers and the end and bridged atoms of the halogen), relationships between intensities in the multiplier, (determination of the number of the "chemically equivalent" atoms of the halogens).

By means of the NQR spectroscopy on the nuclei of ^{35}Cl and ^{79}Br the formal degrees of oxidation of metals and chalcogens were established and approximate forms of their coordination polyhedrons were suggested. The general view of the NQR spectra makes it possible to discriminate two to three major regions of resonant frequencies: especially the high-frequency region characterizing the resonance of the nuclei of the halogens bonded with chalcogens and the low-frequency region characterizing the resonance of the halogens bonded with metals, and in some cases - the intermediate region characterizing the bridged atoms of halogens bonded both with metal and chalcogen (See Table 13). The obtained results suggest that metals and

Table 13. NQR frequencies of 35Cl and 79Br (77K) in chalcohalide complexes of platinum metals and gold [27, 36]

Nos	Compound	, MHz		Nos	Compound	, MHz	
1	PdCl₂(SCl₂)₂	41,031	Cl(with S)	6	PdCl₄(SeCl₄)₂	36,779	
		41,345				36,189	
		18,467	Cl(with Pd)			35,957	
2	PdCl₂(SeCl₂)₂	37,417	Cl(with Se)			35,685	Cl (with Se)
		36,654				35,019	
		17,353	Cl(with Pd)		PdCl₄(TeCl₄)₂	34,661	
3	PdBr₂(SeBr₂)₂	304,06	Br(with Se)			34,520	
		131,169	Br(with Pd)	7			
		32,140	Cl(with Te)		PtCl₄(SCl₂)₂	29,804	
4	PdCl₂(TeCl₂)₂	32,034				29,327	Cl(with Te)
		18,470	Cl(with Pd)			27,132	Cl
5	PdCl₄(SCl₂)s	41,362		8		23,013	Cl(with Pd)
		41,068	Cl(with S)			42,680	
		41,068				42,139	Cl(with S)
		41,016				41,541	
		23,003				41,472	
		22,045	Cl(with Pd)			29,524	
		21,757				29,347	
		21,593				29,062	Cl (with Pt)
9	PtCl₄(SeCl₂)	38,944	Cl(with Se)	14	RhCl₃(SCl₂)₃	28,607	
		38,004				41,360	
		28,404	Cl(with Pt)			41,034	Cl(with S)
		28,145		15	RhCl₃(SeCl₂)₂	40,606	
10	PtCl₄(SeCl₄)₂	36,613	Cl(with Se)			37,174	
		35,020				36,724	
		25,953				36,358	Cl(with Se)
		25,501	Cl(with Pt)			34,919	
11	PtBr₄(SeBr₂)₂	302,830	Br(with Se)			30,061	Cl Se-Cl-Se
		283,102				25,752	
		199,051	Br(with Pt)			20,150	Cl Se-Cl-Rh
		193,352				18,080	Cl(with Rh)
12	PtCl₄(TeCl₄)₂	30,570		16	RhCl₃(TeCl₄)₃	28,511	
		29,830	Cl(with Te)			28,172	
		29,369				27,981	
		27,111	Cl			27,532	Cl(with Te)
		26,939				27,429	
		25,926	Cl(with Pt)			27,382	
13	PtBr₄(TeBr₄)₂	235,050	Br(with Te)	17	IrCl₃(SCl₂)₂SCl₄	42,015	
		218,100				41,125	
		201,450	Br (with Pt)			40,835	
		196,750				40,205	Cl(with S)
		24,542	Cl			39,945	
		22,064	Cl(with Ir)			39,593	
		18,720		21	OsCl₄(SCl₄)₂	27,300	
18		37,648				27,148	
		37,300		22	OsCl₄(SeCl₄)₂	41,864	
		35,463	Cl(with Se)			41,508	Cl (with S)
		34,603				37,352	
		34,560				36,963	

Table 13. Continued

Nos	Compound	, MHz		Nos	Compound	, MHz	
		22,469				16,890	Cl(with Os)
		22,053	Cl(with IR)			16,777	
19	$RuCl_3(SeCl_2)_2$	37,140		23	$OsCl_4(TeCl_4)_2$	30,246	
		36,829	Cl(with Se)			29,719	Cl(with Te)
		36,197				29,210	
		34,805				18,210	
		17,078				18,210	
		16,882	Cl(with Ru)			17,941	Cl(with Os)
20	$RuCl_4(TeCl_4)_2$	28,819				17,460	
		28,545		24	$AuCl_3(SCl_2)$	43,233	Cl(with S)
		28,488				42,547	
		28,141				30,882	Cl(with Au)
		27,969	Cl(with Te)			29,332	
		27,523				25,963	
		27,405				252,976	Br
		27,342				210,666	
25	$AuCl_3(SeCl_2)$	38,988				209,785	Br(with Au)
		37,300	Cl(with Se)	29	$AuCl_3(TeCl_4)$	182,684	
		28,790				31,678	
		28,650	Cl(with Au)			29,810	
26	$AuCl_3(SCl_4)$	42,743				29,318	Cl(with Te)
		42,321	Cl(with S)			29,505	
		41,295				27,808	
		30,613	Cl			27,042	
		27,956		30	$AuBr_3(TeBr_4)$	25,877	Cl(with Au)
		27,917	Cl(with Au)			243,914	
		25,247				239,727	Br(with Te)
27	$AuCl_3(SeCl_4)$	37,751				222,088	
		37,498	Cl(with Se)			211,562	
		37,059				203,935	Br(with Au)
		32,670	Cl				
		29,318					
		26,980	Cl(with Au)				
		26,923					
28	$AuBr_3(SeBr_4)$	311,170					
		310,895	Br(with Se)				
		304,479					

chalcogens in the complexes synthesized in these unhydrous media are in the following degrees of oxidation: Ru (III, IV); Rh(III); Pd(II,IV); Au(III); Os(IV); Ir(III); Pt(II,IV); S(II,IV); Se(II,IV); Te(II,IV).

This makes it possible to identify not only the composition of the ligands , but the composition of the complex as a whole (See Table 13). In some cases presented in the Table, the region of the NQR frequencies of the bridged atoms of the halogen is revealed that occupies on intermediate position in the spectrum, thus inviting the suggestion that the heteronuclear complexes of noble metals and chalcogens have been formed. However, such questions as the methods of mutual coordination, the atoms being coordinated, the symmetry of the coordination site of metal and chalcogen and others remain to be clarified, which was the main reason why the vibration spectroscopy was called on.

The vibration spectra (IR, Raman) were extensively used for proving an individuality of new compounds, identifying the reaction products and studying the structure of the obtained complexes (see Tables 14, 15). The presence of the full vibration spectra consistent with those suggested by the theory of groups, the comparison of the frequencies active in the IR and Raman spectra, the elements displacement (the atoms of the chalcogen and the halogen) in the similar complexes enable the researchers to make well-grounded conclusions about their structure.

The composition and the structure of the complexes vary with the degree of oxidation of the metal and chalcogen. For instance, Pd(II) coordinates only the dichloride form of a chalcogen, while coordination of $\ni Cl_4$ is probably prevented by redox reactions, the coordination saturation of bonds in the formed complexes with $\ni Cl_2$, etc. The complexes M(III) and M(IV) can coordinate both $\ni X_2$ and $\ni X_4$, with the former more often being dichloride and the latter, tetrachloride forms.

This is accounted for by the different conditions under which the complexes are synthesized: M(IV) are stabilized in stronger oxidative media where the most probable form in which chalcogen chloride occurs is $\ni X_4$.

This is also in agreement with the earlier-mentioned "Charge-controlled" and "orbital-controlled" types of their interaction, respectively. Gold (III) and platinum (IV) coordinate both of them.

All the synthesized complexes of the well-studied noble metals (except for Pd(II) and Au(III), are, as a rule, characterized by the formation of a quasioctahedral type of the coordination site of metal. The regular octahedron is found for the complex $[OsCl_6]$ $(SCl_3)_2$. In the cases of Pd(II) and Au(III) the disturbed square-planar coordination sites are formed.

Use of the haloidonor solvents (SCl_2, $SOCl_2$, Br_2) and temperatures <170°C results in the formation of the complexes predominantly with chalcogen tetrahalides. However, use of more indifferent solvents (S_2Cl_2, Se_2Cl_2, CH_2Cl_2) and an increase in the temperature in excess of 170°C contributes as a rule to the coordination of chalcogen dihalides.

In the former case the chalcogen tetrachlorides are coordinated with metals through the bridging atoms of chlorine, but in addition are capable of forming their own quasi-octahedric coordination site (which is especially clearly for Se(IV) and Te(IV)) due to the additional coordination of the atoms of chlorine bonded with the metal, i.e. both NQR, IR and Raman spectra show in this case the formation of heteronuclear complexes with bridging bonds through the atoms of chlorine. In the latter case, coordination of chalcogen and through the chlorine bridge emerging as a result of interaction of the chalcogen with the atom of chlorine of the coordination site of the metal. In these cases, in order to explain the total spectroscopic IR and Raman pattern the authors resort to the idea of forming the heteronuclear complexes of metals and chalcogens. But unlike the preceeding type of the heteronuclear complexes, here the bonds between the quasioctahedric nodes of metals and chalcogens are formed both through the bond M-э and M-X (where э is S, Se, Te, X is Cl, Br).

In some complexes such as Ru(II), Rh(III), Ir(III) there is a more complex picture. For instance, the IR and Raman spectra of the complexes Ru(III) and Rh(III) do not coincide with selenium dichloride which points to the alternative prohibition and the presence of the center of symmetry.

At the same time, unlike the dimeric forms, the monomers $MCl_3\lozenge(SeCl_2)_2$ cannot have the center of symmetry. Besides, at any symmetry the molecules of Ru(II), (d^5 - electron configuration), have an unpaired electron and the complex turns out to be diamagnetic, which is also accounted for by dimerization. Since no difference in the position of the frequencies of the bonds M-Se is found in the vibration spectra of these compounds, while the range of frequencies of valent vibrations M-Cl is found to widen, it is logically enough to suggest the dimeric structure of these heteronuclear complexes with a few bridging atoms of chlorine.

The results of *X-ray crystal structure analysis* obtained by us is a final confirmation of the heteronuclear structure of synthesized complexes of noble

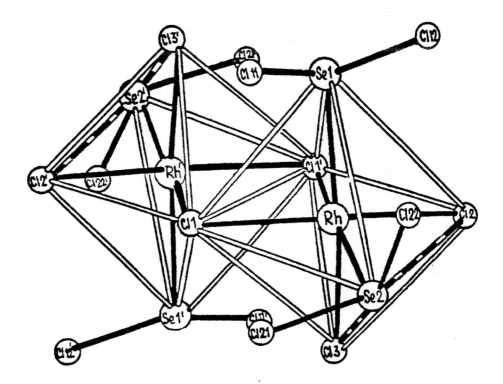

Figure 8. Structure of complex $RhCl_3(SeCl_2)_2$ from X–ray-structure data.

metals and chalcogens. In all the cases the complexes $[Pd(Cl_4)(TeCl_4)]$, $[PtCl_4(TeCl_4)]$, $[OsCl_4(SCl_4)_2]$, $[OSCl_4(SeCl_4)_2]$, $[OSCl_4(TeCl_4)_2]$, $[RhCl_4(SeCl_2)_2]_2$ $[IrCl_3(SCl_2)_2 SCl_4$ analyzed by this method were found to have the *bridging atom of chlorine between two* coordination polyhedrons of platinum metals and chalcogens. As an example, Figure 8 shows the structure of the most sophisticated complex $RhCl_3(SeCl_2)_2$ that is crystallized in monoclinic singony with the parameters: a=7.883 (4)Å, b=10.672 (7)Å, c=12.928 (4)Å, g=95.92 (4)°, r calculated=3.125 g/cm^3, r experimental=3.09 g/cm^3, Z=2, spatial group P2 1/n.

In this centrosymmetrical heteronuclear dimer the atom of rhodium is surrounded by four atoms of chlorine arranged over the distorted octahedron and two atoms of selenium arranged in cis-positions owing to which the angle Se-Rh-Se is greater than the direct one by 18 deg. The peripheral coordination cites of selenium are formed in addition due to the interaction of the neighboring atoms of chlorine, (from rhodium), the distance to which is within the limits of the lengths of the bonds Se-Cl$_{bridging}$ as a result of which the angle Se-Re-Cl decreases by 13 deg [36].

Presently, the X-ray crystal structure analysis of the similar complexes $[PdCl_2(SCl_2)_2]$, $[PdCl_2(SeCl_2)_2]$, $[PdCl_4(SCl_2)_2]$, $[PtCl_4(SeCl_2)_2]$ and others are under way.

Thus, the ideas about the specific character of formation of the adducts, and heteronuclear complexes in such typical solvents with low dielectric permeability as $S2Cl_2$, SCl_2 and $SOCl_2$ suggested in the preceding section on the chemistry of coordination compounds, have found their unambiguous confirmation in the chemistry of the new class of halogen-chalcogenide coordination compounds of platinum metals and gold.

As an example, the new class of the synthesized halogen-halide complexes of noble metals has made it possible to disclose a number of interesting peculiarities of their formation, (their capability to form heteronuclear complexes, lability of the composition of the coordinated forms of chalcogen halides, changes in the donor center in the ligand), confirm the coordination inertness of monohalide forms of chalcogens, establish the series of coordination activity of di- and tetrahalides of chalcogens, etc.

The lability of the composition of coordinated chalcogen halides and changes in their donor atom can be illustrated by the example of synthesis of various chalcogen-halide complexes of gold depending on the composition of the anhydrous solvent and temperature:

$$[AuCl_3 (э Cl_2)] \xleftarrow{CH_2Cl_2} AuCl_3 + э Cl_2 \xrightarrow{SOCl_2} [AuCl_3 (э Cl_4)]$$

$$\uparrow \underline{\qquad\qquad S_2Cl_2 \text{ melt} \qquad\qquad} \uparrow$$

$$SCl_2, SOCl$$

The coordination inertness of monohalides of chalcogens is evidenced by the synthesis with their participation: chalcogen dichloride is coordinated in the case of formation of the complexes $[AuCl_3(SCl_2)]$, $[Os_2O_4Cl_2(SCl_2)]$ and other complexes.

For an analysis of changes in the distribution of electron density in case of formation of complexes with SCl_2 and $SeCl_2$ the calculation of U_p - the number of unbalanced electrons on the atoms of chlorine were conducted. In the platinum complexes $PtCl_4(϶Cl_2)_2$, U_p increases on the equator atoms as compared with the anion $PtCl6^{2-}$.

If $\dfrac{N_x + N_4}{2}$ is constant, the increase in the value of U_p is governed by a decrease in the population N_z of the electrons of chlorine, which in turn results in a more intensive transfer of electron density due to $Cl_{cis}ÆPt$. Thus, replacement of the two chlorine-acido ligands in the trans-position stimulates the transfer of electrons from the cis-chlorine-ions to the atom of platinum, which compensates for weaker-donor properties of the neutral molecules $϶Cl_2$.

In the formation of the $PdCl_2(϶Cl_2)_2$ complexes the number of unbalanced electrons on the atom of chlorine in the cis-bond Pd-Cl remained practically unchanged, though U_p on the atoms of chlorine inside the ligand

	$[PtCl_4(SCl_2)_2]$	$PdCl_4(SeCl_2)_2]$	$[PtCl_6]_2-$	$[PdCl_2(SCl_2)_2]$	$[PdCl_2(SeCl_2)_2]$	$[PdCl_4]_2-$
UP M-Cl	0.53	0.52	0.48	0.32	0.34	0.33
Up ϶-Cl	0.77	0.71	0.72(SCl_2)	0.76	0.69	–

has increased which points to the formation of the coordination bond Pd-϶. These specific features of electron construction of palladium complexes are accounted for by the p-active charge transfer from the atom of palladium to the free low molecular orbitals of ligands. The acceptance of electrons by ligands $϶Cl_2$ in turn accentuates their donor properties, suppresses the transfer $Cl_{cis}ÆPd$ and results in minor changes of U_p on these atoms. The anomalous variation of the NQR frequency of ^{35}Cl with temperature confirms the presence of p-dative interaction.

The X-ray Kb_1-spectra of sulphur and chlorine is an experimental support of the p-dative bonds. The distribution of the population of the p-electrons of sulphur in the molecular r orbitals has been determined by measurements of ratios between the integral intensities of the complex spectra. Comparison of the Kb_1-spectra of the molecule of SCl_2 with the complex $[PdCl_2(SCl_2)_2]$ yields the following results: decreases the population of $C(2b_1 HOMO)$ having the p-antibonding character with a great contribution of the 3-p orbitals of sulphur due to the transfer of electrons on the atom of palladium and the additional short-wave components A, B appear as a result of filling of $LOMO4b_1$ and $5a_1$ of the coordinated molecule of SCl_1 (due to the reverse transfer of electrons from palladium, whereas the population

	A	B	C	ASMO
SCl_2	–	–	1.49	2.18
$[PdCl_2(SCl_2)_2]$	0.38	0.36	0.47	1.95

of all the deeply located MOs does not change.

The parameter U_p is used for calculating the ionicity of the bond M-Cl in the metal complexes of the d^6- and d^8-electron configurations, (provided that there is no d-p contribution to the bond M-Cl), and the charge on the central atom of metal for complexes: +2.3b $PtCl_4$ $(SCl_2)_2$; +1.7b $AuCl_3(SCl_2)$; +0.7b $PdCl_2(SCl_2)_2$.

The conducted analysis of the in-sphere changes in halogen-chalcogenide complexes of noble metals has made it possible to characterize chalcogen dihalides as weak s-donors and moderate p-acceptors.

3. SOME SOLVATO-METALLURGICAL ASPECTS OF PROBLEM

Taking the modern requirements of water-free and energy-and resources-saving technologies into account, which are also waste-free, close-loop and complex, the changeover from pyrometallurgy and hydrometallurgy to solvatometallurgy can be regarded as highly promising. From this viewpoint the use of liquid chalcogen halides as anhydrous solvents possessing both the properties of complex formers and halogenating (primarily, chlorinating) reagents-oxidazers is very promising. They enable the chlorinating process to be conducted at temperatures as low as 80 to 200°C, making it possible to attain as high a concentration of active chlorine as 1000 g/1, which in unit volume is 270 to 370 times as high as in gas, (in the bonded state), allow the halogenating processes to be transformed from low heterophase into a "low-temperature" rapid homophase, making it possible to limit the use of highly-toxic chlorine making processes less ecologically harmful, etc.

And taking the possibility of realization of a series of reactions into account which can schematically be represented as:

thermal decomposition $[MX_n(э X_m)_p] \xrightarrow{\sim 200°C} MX_n + p \, э \, X_m \xrightarrow{>200°C} M +$

exchange $[MX_n(э X_4)_p] + M \, э'X_4 \xrightarrow{Solv} [MX_n(э'X_4)_m + p(э X_4)$

reduction $[MX_n(э X_m)_p] + H_2 \xrightarrow{>700°C} M + э + H_2 \, э + HCl$

hydrolysis $[MX_n(э X_m)] + H_2O \longrightarrow MX_n(OH)_m^{m-} + э + H_2 \, э)_3 + HX$

they open additional opportunities of solvatometallurgy and technology.

IV. CHEMISTRY OF COORDINATION COMPOUND SIN THE VAPOR (GAS) PHASE

Until recently, the chemistry of coordination compounds in the vapor (gas) phase has been an insufficiently studied field of science, its development is stimulated both by scientific and practical interests.

The interest in the field has been shown by theoretical scientists and experimenters are accounted for by the specific features of the phase state of these compounds characterized by the absence of the effects of interaction with their own crystal lattices or interaction with the molecules of the solvent, and at the same time, by the presence of coordinately-unsaturated bonds, (with resultant dynamic non-rigidity of bonds), thermally-excited particles and other effects. In the opinion of specialists in applied chemistry, conversion of some coordination compounds into the gas phase makes it possible to conduct the gas-transport reactions and to obtain water-free wasteless and highly efficient technologies for the production of metals and coatings with valuable properties.

Chemistry of coordination compounds in the vapor (gas) phase is represented chiefly by a few types of volatile complexes, (with the compounds of double-salt type, for instance, halides, (Alk_nML_m), are excluded here from the list), presented in Table 16. Extensive studies of these compounds began in the 1970's, though a number of them, primarily carbonyls and cyclodienyls have long been known.

In the author's opinion, the problem of purposeful formation and utilization of volatile complexes with desired properties-high volatility, thermal stability and controlled thermal decomposition paths, can be solved only through the detailed study of the electron and the geometrical structure of their individual molecules with further investigation of the specific features of their intermolecular interaction ensuring conversion into the vapor phase and researches into the thermodynamics of this process and, at last, into the mechanism and kinetics of their thermal transformations. That is why in this section of the paper the authors would like, on the example of the class of the easily volatile b-diketonate complexes of metals actively synthesized and studied, their N- and S-containing analogs, fluorinated derivatives of b-diketonates and their adducts with various donor molecules (Table 17) to dwell upon the following important problems of the chemistry of coordination compounds in vapor (gas) phase [35-37]:

- quantum chemistry, spectroscopy and structure of volatile b-diketonates in the gas-phase, including the problem of dynamic non-rigidity of bonds in the chelate cycle;
- problem and major criteria of volatility of such coordination compounds;
- gas-transport, homophase, heterophase processes with participation of high-volatile b-diketonate complexes.

Table 16. Basic types and characteristics of volatile coordination compounds.

Nos	Ligands	Metals	General formula of complex	Research methods	Temperature(°C): boiling (m), melting (n), sublimation (s), chromatography (ch), vapor pressure, mmHg(p)
1	2	3	4	5	6
1	b-diketones	p-, d-, f- metals		TA, MS, ED, GC, RS, EAS	t_m=20-200 t_b=40-350 p=10-200
2	Thio-derivatives of b-diketones	Ni^{II}, Co^{II}, Zn^{II}, Cd^{II}, Pd^{II}, Pt^{II}, etc.		TA, GC, MS	t_m=120-280 t_{ch}=110-300
3	b-ketomines	Cu^{II}, Ni^{II}, Pd^{II}, Pt^{II}, etc.		TA, GC	t_mª200 t_{ch}·120ª280
4	Oxyacids (acetates, propionates)	Be^{II}, Zn^{II}, etc.	$M4O(R)_6$	TA, GC	t_{ch}ª150 (decomp)
5	Dialkyl-thiophosphates	Au^{I}, Cd^{II}, Zn^{II}, etc.		GC	t_{ch}ª200 (decomp)
6	Xanthogenates	Co^{II}, Ni^{II}, Pd^{II}, etc.		TA	t_s=70-100° p·10^{-3}
7	Carbonyls	Fe^{III}, Ni^{II}, Mo^{VI}, W^{VI}, etc.	$M(CO)_m$	TA, MS, ED, IRS	t_m=80-150 t_b=100-300
8	Cyclodenyls	Fe, V, Mo, Mn, Cr, etc.		TA, EG	-70-280 =10^{-2}-120

Table 17. Basic b-diketones and metal complexes with them synthesized by the authors

R$_1$-COCH$_2$CO-R$_2$		Abbreviation	Synthesized complexes
R$_1$	R$_2$		
CH$_3$	CH$_3$	NAA	Be, Al, Ga, In, Sc, Ti (III, IV), V (III, V), Cr (III), Fe (II, III), Co (II), Ni(II), Cu (II), Zn (II), Zr (IV), Hf (IV), Nb(V), Ta (V) Mo (III, V, VI), W (III, VI)
CH$_3$	CF$_3$	NTFA	Be, Al, Ga, In, Sc, Ti (III, IV), V (III, V), Cr (III), Fe (II, III), Co (II), Ni (II), Cu (II), Zn (II), Zr (IV), Hf (IV), Nb (V), Ta (V), Mo (III, V, VI), W (VI)
CF$_3$	CF$_3$	NGFA	Al, Ga, In, Ti (III, IV), V (III, V), Cr (III), Fe (II, III), Co (II), Ni (II), Cu (II), Zn (II), Zr (IV), Hf (IV), Nb (V), Mo (III, V, VI), W (III, VI)
C(CH$_3$)$_3$	C(CH$_3$)$_3$	HDRM	Al, Ga, In, Cr (III), Fe (III), Co (II), Ni (II), Cu (II), Zr (IV), Hf (IV)
C(CH$_3$)$_3$	CF$_3$	NPTFA	Al, Ga, Ti (IV), Fl (III), Cu (II), Zr (IV), Hf (IV), Nb (V)
CH$_3$	C$_3$F$_7$	NGFG	Al, Ga, In, Ti (III, IV), Cr (III), Fl (III), Co (II), Cu (II), Zr (IV), Hf (IV)
C(CH$_3$)$_3$	C$_3$F$_7$	NFOD	Al, Ga, Cr (III), Fl (III), Co (II), Ni (II), Cu (II), Zr (IV), Hf (IV)
CH(C$_2$H$_5$)$_2$	C$_3$F$_7$	NGFEN	Al, Cr (III), Ti (IV), Cu (II), Zr (IV), Hf (IV)
CF$_3$	C$_3$F$_7$	HDOG	Al, Cr (III), Cu (II), Zr (IV)
CH$_3$-C(NH)	CH$_2$CO-CH$_3$	NNAA	Co (II), Ni (II), Cu (II)
CH$_3$-C(NH)	CH$_2$CO-CF$_3$	NNTFA	Ni (II), Cu (II)
CF$_3$-C(NH)	CH$_2$CO-CF$_3$	NNGGA	Ni (II), Cu (II)
C(CH$_3$)$_3$-C(NH)	CH$_2$CO-CF$_3$	NNPTFA	Ni (II), Cu (II)
CH$_3$-CO-CG=	CSH-CH$_3$	NMTA	Cu (II), Mo (III, V), W (III)
CF$_3$-CO-CH=	CSH-CH$_3$	NTFNTA	Mo (III, V), W (III, V)
CF$_3$-CO-CH=	CSH-CF$_3$	NGHMTA	MO (III,V), W (III, V)

1. QUANTUM CHEMISTRY, SPECTROSCOPY AND STRUCTURE OF VOLATILE B-DIKETONATE METALS AND VAPOR (GAS) PHASE [38,41,44,47,49,50,54]

a) Nature of chemical bonds and stability of molecules of b-diketonate metals.

The analysis of the orbital state of the p-diketonate complexes of the metals studied shows that, in all cases, from the viewpoint of the ratio of the population of bonding, nonbonding, and antibonding molecular orbitals are essentially stable compounds.

All the valence electrons of the metals of group II may be arranged on the bonding levels of the complex M(b-dik)$_2$ of symmetry T_d (capacity of the bonding orbitals is 12 electrons), in the subgroup of complexes of III A elements, (Al, Ga, In) M(b-dik)$_3$ at octahedral configuration of the surrounding ligands all the bonding orbitals, (a_{1g}, t_{1u} and t_{2g}), are populated and only 4 electrons are on the non-bonding orbital t_{1g} belonging to the b-diketone.

The levels of quasi-octahedral complexes of the 3-d metals M(b-dik)$_3$ gradually fill with the nonbonding t_{1g} and t_{2g} orbitals of the b-diketone, with all the bonding orbitals of the complex being filled, and only in the last representatives of the series of the 3-d metals (Ni, Cu, Zn) the antibonding orbitals t_{2g} begin to be filled. Hence, the highest stability should be expected in the complexes with the vacant and filled nonbonding orbital t_{1g}, (Sc (III) and Co(II) and in the complexes Cr (III), with a half-filled orbital with maximum multiplicity (t_{1g}^6).

This is in agreement with the stability experimentally observed during thermal decomposition of the b-diketonates of the series of 3d-metals with coordination number 6: Sc$\gg$Ti$>$V$<$Cr$>$Mn$\leq$Fe$>$Co.

The analysis of the general schemes of filling the orbitals of the p- and d-metals makes it possible to make the conclusion of the p-metals, (oxidation state - three), tends to form stable complexes with a maximum coordination number, i.e. coordinately-saturated compounds. The d-metals, (oxidation state - two), predominantly form coordinately-unsaturated compounds of the T_d or D_{4h} symmetry with the expressed ability to attach additional ligands either through forming adducts or by way of their oligomerization. In all the cases, the attachment of the ligand occurs with the interaction of its orbitals with the nonbonding frontier orbitals of the complex at configurative transformation of the initial T_d symmetry of the complex into O_h (or D_{4h}) (interaction with the nonbonding orbital a_{1g} in O_h or a_1 in D_{4h} localized on the atom of the metal).

In order to clear out the specific features of the electronic structure of the b-diketonate complexes of the metals, the contribution of each atom of the 6-member chelate ring to the system of delocalized bonds of the complex and the effect of the d-substitutes on the transfer of electron density from the central atom to the ligand, calculations have been conducted for a number of complexes of p- and d-metals with due account taken of all the chelate atoms including the a-substitutes by the PPDP/2 method on the basis of up to 232 orbitals.

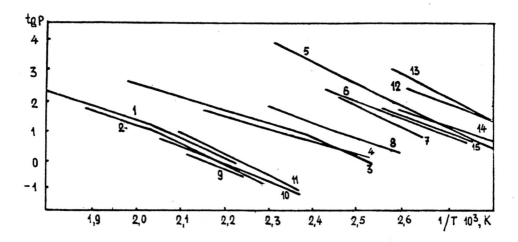

Figure 9. Relative positioning and designations of atoms in B-diketonates.

The comparative analysis of reactionary-labile highest occupied molecular orbitals (HOMO) and the lowest unoccupied molecular orbitals (LUMO) with the use of the calculated diagrams of three-dimensional distribution of electron density has shown the delocalized character of the HOMO complexes of various partial-two center type: in terms of bond M-O of bonding type with predominant s-component; in terms of bonds C_1-O and C_1-C_3 of the s-bonding and pi-antibonding type with the p-component less than the p-contribution to the bond M-O; and the nonbonding component in the field of the bond C_1-C_2 (Figure 9).

Table 18 shows the values of the full, (within the framework of the method PPDP/2), electron population on the central atom and on the bonds in the complexes of aluminum (III), chrome (III) and nickel (II).

In the complexes of metal as compared with hydrogen (malonaldehyde) the CH_3-group is the electron-donor, while the CF_3-group is the electron-accepting. The values of the added electron density on the central atoms of metal and on bonds in the series of various a-substitutes for both p- and d-metal point to the relative independence of the chains of the inductive, (in terms of s and p-components), effect on the conjugate chain C_3-C_1-O-M and C_1-C_2-C_1^1 of the chelate.

The nature of the central atom has considerably smaller effect on the change of electron density of various components of the six-member ring, i.e. though there is the effect similar to the inductive effect of the a-substitutes, this effect is much weaker. This effect appears to be governed by the ability of the metal to give off or accept the electron density and is generally well correlated with the values of the Malliken's electronegativity.

Table 18. Electron population on central atom in bonds of some B-diketonates of metals

Complex	Atom or bond in complex	R_1=H (MA)	R_2=H	R_1=CH$_3$ (AA)	R_2=CH$_3$	R_1=CH$_3$ (TFA)	R_2=CF$_3$	R_1=CF$_3$ (HFA)	R_2=CF$_3$
		$q_{ОБШ}$	q	$q_{ОБШ}$	q	$q_{ОБШ}$	q	$q_{ОБШ}$	q
AlL$_3$	Al	2.258	0.722	2.316	0.741	2.180	0.685	2.040	0.628
	Al-O	3.001	0.510	3.069	0.523	2.825	0.484	2.581	0.443
	O-C$_1$	2.831	0.610	2.905	0.626	2.687	0.579	2.464	0.531
	C$_1$-C$_3$	3.269	0.219	3.359	0.225	3.105	0.208	2.851	0.191
	C$_1$-C$_2$	2.816	0.609	3.862	0.607	2.776	0.608	2.633	0.601
CrL$_3$	Cr	5.601	1.803	5.744	1.850	5.407	1.711	5.059	1.568
	Cr-O	2.115	0.360	2.164	0.369	1.932	0.341	1.820	0.313
	O-C$_1$	2.896	0.602	2.911	0.618	2.692	0.572	2.469	0.524
	C$_1$-C$_3$	2.992	0.176	3.074	0.182	2.841	0.167	2.611	0.154
	C$_1$-C$_2$	2.870	0.575	2.917	0.573	2.892	0.574	2.684	0.568
NiL$_2$	Ni	9.462	4.126	9.705	4.235	9.135	3.915	8.548	3.589
	Ni-O	2.648	0.326	2.708	0.334	2.493	0.309	2.277	0.283
	O-C$_1$	2.932	0.615	3.009	0.631	2.783	0.584	2.552	0.535
	C$_1$-C$_3$	2.778	0.218	2.854	0.224	2.638	0.207	2.405	0.190
	C$_1$-C$_2$	3.283	0.611	3.337	0.609	3.237	0.610	3.070	0.603

The laws obtained by quanto-chemical calculations were supported by the experiments for measurements of bond energy of the core-shell electrons in b-diketonates of aluminum, gallium and indium which were conducted by us with the use of the X-ray photoelectron spectroscopy and showed a decrease in the electron density on the atoms of the metal in the series of the ligands: NAM, NTFA, NGFA, i.e. in proportion with the s*-induction constants of the a-substitutes in b-diketone.

The results of the quanto-chemical calculations and X-ray photoelectron spectroscopy make it possible to conclude that, in the b-diketonates of metals, distribution of electron density is delocalized in character both over the ligand skeleton chain and the atom of metal and over the chain connecting the a-substituent through the C-O group with the atom of metal. These two delocalized systems can practically be regarded as independent which to a considerable extent distinguishes the b-diketonate complexes from the classical organic heterocyclical systems.

Non-uniformity of electron distribution in the molecule and inclusion of substituents into the delocalized chain -M-O-C1-C3(R) points to the absence of the quaziaromaticity of the chelate ring, though the presence of the delocalization in the six-member chelate ring results in its vague similarity, for instance, in the possibility to perform the nucleophile substitution reaction in the (C2) carbon atom of the chelate ring.

Thus, the quanto-chemical studies of isolated molecules of the b-diketonate complexes establish the conjugation between the components of the chelate chain, different type of each of the two-center components of this chain in complexes and inductive effect of the a-substitutes on the distribution of electron density in complexes, i.e. the basic factors responsible for specific chemical and physical properties of b-diketonates of metals.

The sufficiently high thermal stability (50-250°C) and volatility (13.3-26600 Pa) of the b-diketonate complexes of metals permit them to be studied by the electron absorption spectra (EAS) and IR-spectroscopy methods in the gas phase for which purpose the special high-temperature attachments to spectrophotometers have been designed and the techniques have been worked out for recording the EAS and IR spectra within the ranges of temperatures of 50 to 400°C and frequencies of 50000 to 400 cm^{-1}. The use of these attachments and techniques have made it possible to establish the composition and construction of b-diketonate complexes and adducts of the series of 3d-metals in the gas phase.

b) Electronic absorption spectra (EAS) of b-diketonates of 3d-metals and their adducts [38, 39, 45, 46]

For the first time, the EAS were taken and analyzed for the synthesized b-diketonate complexes of metals: Cr (III), Fe (II, III), Co (II), Ni (II), Cu (II) with the b-diketones: NAA, NTFA, NGFA, NDFG, NGFG, NGFEN, NNAA, NNTFA, NNGFA, NMTA and adducts with dimethylformamide (DME), pyridine (P$_y$), methanol, ethanol (alk), tributylphosphine oxide in the gas phase and for comparison - in solutions of organic solvents or in solid phase. The analysis of EAS was conducted with the use of the ligand field theory.

For all the investigated complexes the transfer to the gas phase results in the shift of the d-d absorption strips to within the long-wave range (bathochromic shift) which is caused by both the weakening of the ligand field with an increase in temperature and by filling higher oscillatory sublevels. For the complexes having the center of symmetry it is seen (Figure 10) that the intensity of the strips in the spectra increases with increasing temperature.

Table 19 shows the geometrical characteristics of some b-diketonates of metals. The complexes of chrome (III) both in the gas phase and in melts are characterized by the octahedral type of symmetry (Figure 10, a, b). The two-valent ions of manganese, iron, cobalt and nickel form the complexes with b-diketones with coordination number of six, with such a coordination being accomplished due to the formation of both the oligomers and the adducts with two molecules of additional ligands. The adducts of cobalt, nickel and copper with TFA, GFA and water or alcohol in the solutions of organic solvents are characterized by the octahedral type of the structure with tetragonal distortion (symmetry D_{4h}) in conformity with the number and position of the EAS strips. In the gas phase the molecules of water, (or alcohol), are detached thus causing the coordination number of the ion of metal to reduce to four. When the adduct is formed by the ligand possessing more pronounced donor capabilities, for instance DMFA, in the cases of manganese, cobalt and nickel it does not detach when the complex is transferred to the gas phase (Figure 10b). In the gas phase and solutions, the adducts of two-valent 3d-metals with such donors have tetragonally-distorted (D_{4h}) octahedral structure.

The further increase in intensity of donor properties of additional ligands, (for instance, the introduction of TBFO), results in the formation of complexes with the coordination of the distorted tetrahedron (Figure 10d) with a simultaneous strengthening of its bond with the central atom and with considerably greater redistribution of the electron density in the coordination mode of chelates.

For example, in some complexes the calculated spectroscopic parameters of electronic structure have made it possible for b-diketonates of chrome (III) to arrange the ligands in the following spectroscopic series in decreasing order of the values of parameter 10 D_q in the gas phase: GFEN>TFA> DFG>AA>GFA. The assessment of covalence of the bond in terms of nepheloxetic effect B has shown that the complex of chrome (III) with TFA is characterized by the most conic character of the bond as compared with other b-diketones: TFA (B=0.89)>GFEN>(0.81)>DFG(0.78)>GFA(0.72)>AA> (0.61). The similar series is also true for the adducts of cobalt (II): Co (TFA)$_2$◊2DMFA (0.86)>Co(GFA)$_2$◊2DMFA (0.76). The comparison of the similar complexes of cobalt (II) and nickel (II) shows that the bond in the complexes of nickel is more ionic in character: NiL$_2$ (0.89)>CoL$_2$ (0.76) and the introduction of a stronger donor ligand into the adduct results in an increase of covalence of bonds in the complexes: Co(GFA)$_2$◊2DMFA (B=0.76) Co(GFA)$_2$◊2TBFO(B=0.69).

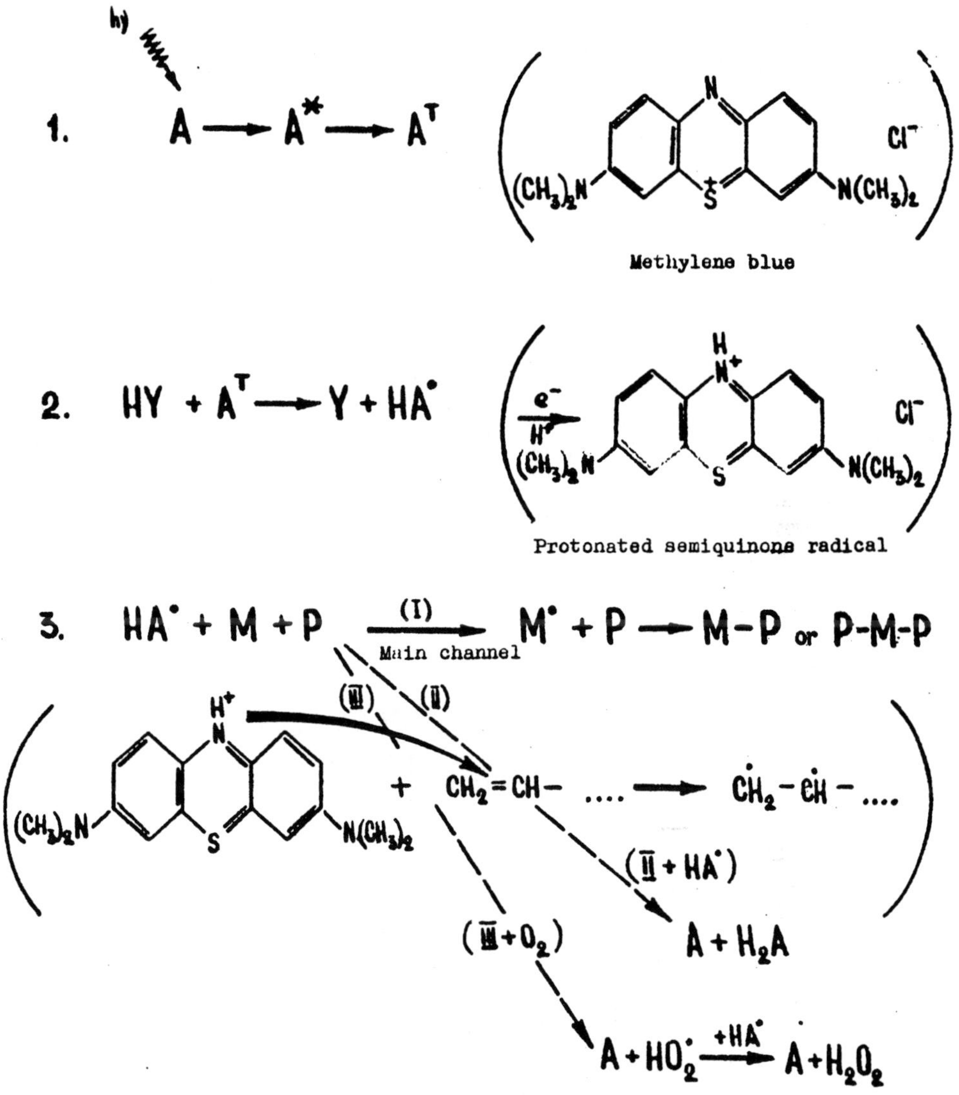

Figure 9. Relative positioning and designation of atoms in B-diketonates

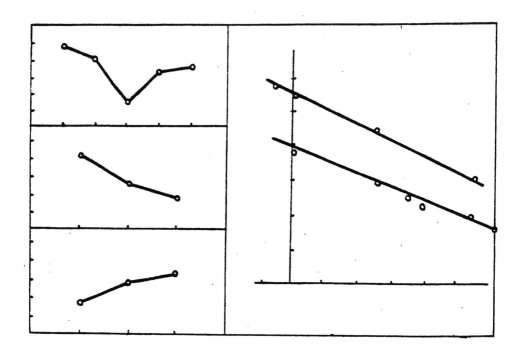

Figure 10. Electron absorption spectra: a - Cr(GFA)$_3$; b) - Cr (GFEN)$_3$; b) - Co (GFA)$_2$x2DMFA; c) - Co(GFA)$_2$x2TBFO (1 - in benzene; 2, 3 - in gas phase).

c) IR-spectra of b-diketonates and their adducts.
The best way to conduct the analysis of IR-spectroscopy of b-diketonate complexes in the gas phase is the discussion of three temperature regions: transfer of the compound into gas phase; stable existence of the complex in the gas; the regions of predissociation of the complex in gas phase.

In transferring from the solid to gas phase the shift of the IR-frequencies is observed which is accounted for by both the change of intermolecular interaction and by the change in the configuration of the complex.

As the temperature increases, the spectra of the complexes in the gas phase exhibit the widening of the strips due to the juxtaposition of the spin components of the spectra, with the greater widening being characteristic for the strips associated with the oscillations of the end groups of the substitutes.

The series of complexes M(AA)$_m$-M(TFA)$_m$-M(GFA)$_m$ (M-Al, Ga, In, Cr (II), Co (II), Ni (II), Cu (II)) retains the same order of the shift of the oscillation strips C-O, C-C towards higher frequencies as compared with the solid phase spectra, which is associated with the positive inductive effect of the CF$_3$ group.

In the series of complexes Al (or Ga, In), L_3 and Co (or Ni, Cu) L_2 (where L-TF and GFA) in the gas and solid phases an increase in the frequency of oscillations of the bond M-O is observed, which is associated with the growth of the electron-accepting capability of metals in these series.

All complexes, particularly those with unsymmetrical ligands (TFA, GFEN, GFG, etc.) in the gas phase are distinguished with a greater number of fixed frequencies as compared with the spectra in the crystalline state that is accounted for by the structural inequality of the M-O and C-O bonds manifesting itself in the gas phase.

On approaching the temperatures of thermal decomposition, practically all the studied complexes exhibit the splitting of the absorption bands containing the valent oscillations of M-O, C-O, C-C bonds which is caused by the distortion of symmetry of the coordination site up to the opening of the chelate ring and appearance of the frequency of oscillation of the free C=O group of b-diketone.

The IR spectra of the adducts $M(GFA)_2 \lozenge 2DMFA$ (M-Min (II), Fe (II), Co (II), Ni (II), Cu (II) in the gas phase taken in the region of their thermostable existence are similar to the IR spectra of the adducts in the solid state. Only minor shiftings of frequencies are observed, caused by a weakening of intermolecular interaction.

At temperatures higher than 200°C in the IR spectra of the adducts in the gas phase the strips begin to appear which increase in magnitude with increasing temperature and which are associated with the oscillations of free bond C=) DMFA, i.e. detachment of molecules of the donor ligand takes place (detachment of molecules of DMFA is also confirmed by the results of gas-chromatographic analysis). The adducts with strong donors of the TBFO type are distinguished by the equilibrium of several forms of the complex due to the bidentate and monodentate coordination of b-diketone.

d) *Dynamic non-rigidity of bonds of the coordination site of b-diketonates and other structure and geometrical peculiarities*

The optimization of the geometry of b-diketonates and b-diketonate complexes conducted by the method of gradiental descent over the hypersurface of their full energy has shown that the b-diketone itself is characterized in the complex by the sterically rigid structure and that all the substantial changes in the geometrical structure of b-diketonates occur only in the coordination site of chelate.

These changes are in full agreement with the partially-bicentral character of bonds which is disclosed in b-diketonate complexes and which results in that as the electron population increases the bonds of binding partially-bicentral character decrease in length, whereas those of non-binding character - increase in length. The calculations have shown that the lengths of M-O and $M-O^1$, C_1-O and C_1-O_1, C_1-C_3 and $C_1^1 - C_3^1$ bonds, respectively are not equal to one another even for the ligands symmetrical in terms of substitutes. For instance, for complexes of aluminum this difference in the lengths of M-O and M-O bonds amounts to: A-Al-O (Å) with (AA): 2.064, with (GFA) 2.090 Al-

0^1(Å) with AA:2.082 with GFA:2.113. The curves of the obtained relationships between the total energy of the complexes, (cross-sections of the total-energy surface by the planes along the line of the metal-ligand bond passing through the center of symmetry of the ligand environment), for Al (AA)$_3$, Al (TFA)$_3$ and Al (GFA)$_3$ and the shift of the atom of the metal are shown in Figure 11.

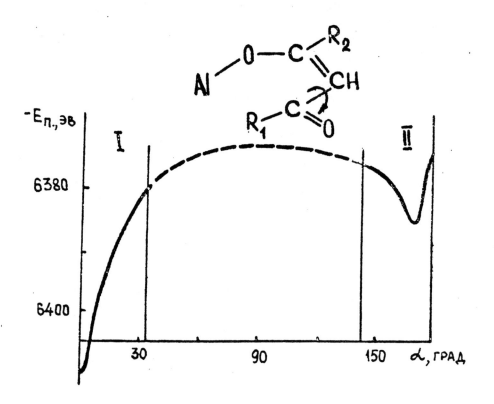

Figure 11.   Potential curves of variations of total energy of molecules of b-diketonates of aluminum.

Each of the potential curves has two minima separated by minor (a few J/mol) barrier, i.e. the full potential surfaces, for instance tris-complexes ML$_3$ in the coordinates of shift of the central atom possess six minima in which at any moment the atom of the metal may be located, and hence, the coordination site of the molecule is asymmetrical as a whole.

Because of small values of the barrier the intramolecular tunnelling of the central atom or anharmonic oscillatory transformations with the frequency of the order of 10^{13} s^{-1} are possible, with the introduction of fluorinated a-substituents and increase in the mass of the central atom resulting in a decrease of the frequencies of these transformations.

The analysis of the variations of the values of splitting the potential curves has shown that the values of shifting and the height of the barriers decrease with the increasing radius of the central atom and increase with growing electron-acceptor properties of the a-substitutes of b-diketone.

This behavior is in full agreement with the concept of dynamic non-rigidity of the coordination node of the molecule, i.e. unlike the sterically rigid ligand chain the coordination node in the b-diketonate complexes of metals possesses dynamic non-rigidity.

The analysis shows that this non-stability is caused by the pseudoeffect of Yan-Teller caused by the interaction of the HOMO and LUMO of the b-diketonates making a considerable contribution to the adiabatic potential of the molecular system.

Experimentally, this instability is confirmed unambiguously by the considerable contribution to the total dipole moments of the b-diketonates of metals and by the splitting of the frequencies of oscillations of M-O bonds in the IR spectra of the complexes in the gas phase, for instance by the appearance of the doublet for Al $(AA)_3$-417 and 427 cm^{-1} and Al $(GFA)_3$-413 and 424 cm^{-1}.

Another specific feature of the structure of the b-diketonates of metals is the possibility of realization of the equilibrium between the complexes containing both the bidentantly- and monodentantly-coordinated b-diketone simultaneously.

The analysis of variations of the full energy of the molecule of b-diketonate, (on the example of tris-2.4 aluminum pentandionate), in the case of the torsional opening of the ring through rotation around the axis of the group passing through the atoms of carbon of the ligand skeleton, (in the b- and g-positions), has shown that the relationship between the full energy of the molecule and the angle of rotation is characterized by two minima (Figure 12). The presence of the first suggests that the energetically most suitable one for the b-diketonates is the bidentant way of coordination of the ligand with formation of the flat chelate ring, which is in full agreement with the generally-accepted structure of the b-diketonates of metals. The parabolic form of the curve in the vicinity of this minimum (a<10°) confirms the harmonic character of the deformative oscillations in the molecule of the b-diketonate, which earlier was only an *a priory* assumption. The second, though not as deep, but clearly an expressed minimum of the full energy at values a·170° confirms the possibility of stable existence of the "open" state of the ring.

Opening of the chelate ring can be expected under both the physical effects: UV, electromagnetic excitation, mass-spectrometrical fragmentation, heating up to the temperatures close to the temperatures of thermal dissociation of complexes and sometimes under chemical effects: formation of mixed ligand complexes (adducts).

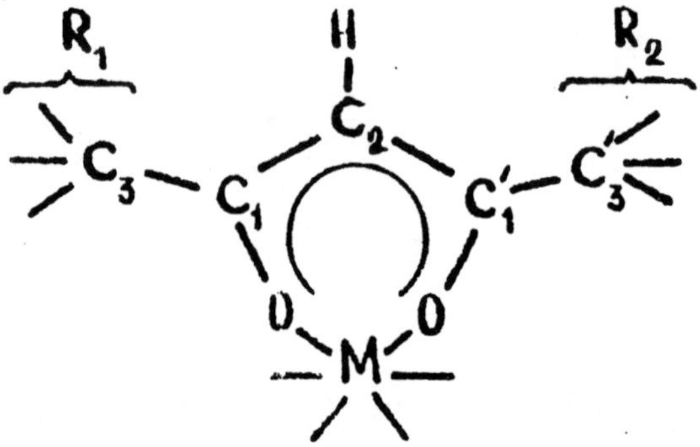

Figure 12. Variation of total energy of molecule Al(AA)$_3$ at rotation of group-C
tantamount to the opening of the chelate ring.

Such transformations in the structures of b-diketonates have been confirmed experimentally. The EAS studies in the gas phase with an increase in
temperature have revealed a considerable distortion of the coordination of
chelates of the 3d-metals (Table 19). The process of opening of the chelate
ring can be seen clearly enough within the temperature interval close to decomposition of complexes in the studies of oscillation spectra. For instance,
for bis-1, 1, 1-trifluor-2.4-pentandionate of cobalt (II) at temperatures above
150°C in the spectrum of gas phase there appears the frequency of oscillations of the free carbonyl group n(C=O) at 1660 cm^{-1} and the intensity of the
strips n(M=O) at 409, 431 and 615 cm^{-1}. The similar picture is also seen for
other b-diketonates of p- and d-metals.

This is easily explained by opening the diketonate ring in a part of the
complexes with an increase in temperature and by the establishment of equilibrium between bidentant and monodentant coordination of b-diketones in the
complexes.

The data obtained by quantum-chemistry and spectroscopic studies makes
it possible to establish the basic peculiarities of the electronic and geometrical structure of individual molecules of b-diketonate complexes of metals and
their behavior in the t-gas phase.

2. PROBLEM AND CRITERIA OF VOLATILITY
OF B-DIKETONATES OF METALS

*a) Intermolecular interaction potentials and their relationship with structure of
b-diketonates*

To disclose the relationships between the intramolecular characteristics
of individual molecules and volatility, stability and specific features of
thermal decomposition of b-diketonates of metals the authors have conducted
the studies of the behavior of the ensemble of their molecules in the pro-

cesses of the transformation: condensed phaseÆgas phaseÆproducts of thermal decomposition.

The possibility of molecular reversible transformation solid body $\rightleftarrows$ vapor is governed by the intermolecular interaction forces depending mainly on the electronic and geometrical structure of b-diketonate complexes.

The direct quantum-chemistry analysis of parameters of the paired intermolecular interaction potential (PIIP) is practically impossible now because of the limited capabilities of analysis techniques. Therefore, the authors have proposed and realized the PIIP analysis model with the use of both the parameters of distribution of electron density in the molecules of b-diketonates, (obtained with the use of the PPDP/2 method), and a number of empirical parameters describing the atom-atom interaction in these systems:

$$U(R) = q_{AB(R)} + W_{AB(R)} = \sum_{iEA}\sum_{jEB} \frac{q_{iA} + q_{iB}}{R_{ij}} + \sum_{iEA}\sum_{jEB} 4(\varepsilon_{iA} + \varepsilon_{jB}) \cdot (d_{ijAB} - d^2_{ijAB})$$

where $U(R)$ is the PIIP; $q_{AB(R)}$ is the electrostatic component of PIIP; $W_{AB(R)}$ is the "non-valent" (van-der-Waal's) component of PIIP; q_{ij} are the values of the charges on atoms obtained by the PPDP/2 method; R_{ij} are the atom-atom distances; d_{ij} is the Van-der-Waals' radius of the 1⁻, j-atom; e_{ij} is the empirical parameter.

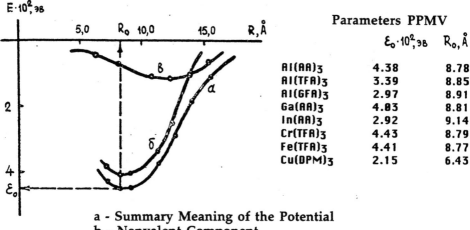

Parameters PPMV		
	$\varepsilon_o \cdot 10^2$, эв	R_o, Å
Al(AA)₃	4.38	8.78
Al(TFA)₃	3.39	8.85
Al(GFA)₃	2.97	8.91
Ga(AA)₃	4.83	8.81
In(AA)₃	2.92	9.14
Cr(TFA)₃	4.43	8.79
Fe(TFA)₃	4.41	8.77
Cu(DPM)₃	2.15	6.43

a - Summary Meaning of the Potential
b - Nonvalent Component
c - Electrostatic Component

Figure 13. Potential curves of intermolecular interaction energies for Al(AA)₃ with their components (a, b, c) and parameters PPMV of some b-diketonates of metals (₀, R₀).

The PIIPs were computed for the p-(Al, Ca, In) complexes and (Cr, Fe, Cu) d-metals with some b-diketones (NAA, NGFA,, NDPM and NGFEN). The form of the calculated orientation-averaged PIIP, for instance for Al $(AA)_3$ (in the form the calculated potentials are equal for all the calculated complexes) consisting of the electrostatic and "non-valence" components is shown in Figure 13 with parameters of PIIP (e_0 and R_0) for some complexes.

The validity of the calculated values of the PIIP parameters have been confirmed experimentally by comparison of the values of viscosity of gaseous b-diketonates of metals calculated from these parameters and the values experimentally obtained with the use of the "capillary-in-gas-chromatographic version" of the method developed.

Calculations have shown that the basic contribution to e_0 is made by the "non-valent" interaction and that the value of e_0 depends considerably on the electronophilic properties of the ?-substitutes in the ligands forming the periphery of the complex.. By shielding the coordination node with metal weakens the interaction between the neighboring molecules, i.e. form the "island"-type structure of the complex. The nature of the central atom affects the value of e_0 to a lesser degree, though it is seen that an increase in electronegativity of the metal results in an increase in the PIIP potential which should lead to a lower volatility.

The analysis of the PIIP parameters of the b-diketonates of both p- and d-metals with the b-diketones different in nature make it possible to draw some general conclusions.

For the complexes with a strongly shielded coordination node, (for instance, coordinately-saturated tris- and tetra-kis-b-diketonates), the dipole-dipole interaction make inconsiderable contribution to PIIP, despite large values of m, and the main contribution is made by the Van-der-Walls forces of interaction between the atoms of a-substitutes forming the molecular periphery of the complex. Replacement of the hydrogen atoms in the ?-substitutes for the fluorine atoms should increase volatility, whereas their replacement for the CH_3 or other alkyl groups (including branched chain a-substituents of the type $C(CH_3)_3$ or - $CH(C_2H_5)_2$), should decrease it.

For complexes with a weaker screened coordination node, (for instance, coordinatively-unsaturated bis-b-diketonates of the 3-d metals (II)), an important part is played by the orientation effects. In the case of complanar arrangement, just as in the case of maximally coordinated complexes the greatest contribution is made by the Van-der-Waals forces of interaction between the periphery atoms. But in the presence of parallelism between the planes of molecules of bis-complexes the interaction of the periphery atoms is insignificant as compared with the values of intermolecular interaction of completely coordinated nodes of complexes which includes the interaction metal-donor atom of the neighboring molecule, etc. Therefore, unlike the maximally coordinated b-diketones in the incompletely coordinated b-diketones volatility may increase not only in the case of replacement of hydrogen atoms for fluorine atoms but in the case of their replacement for branched-chain groups of the $-(C(CH_3)_3)$ or other groups.

b) Thermodynamics of transformation of b-diketonates of metals into the vapor phase

For obtaining the data on the thermodynamic parameters of evaporation and sublimation of volatile b-diketonates the authors studied the saturated vapor pressure as a function of temperature using the statistical method with the membrane null-pressure gauge and the gas-chromatographic method in the flow that was developed. The measurement results have been processed with the use of the special computer program and represented as the equation $\ln p = B - A/T$.

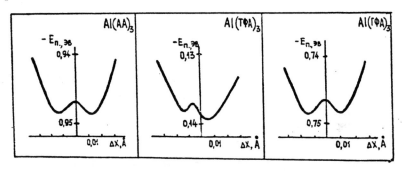

Figure 14. Variation of saturated vapor pressure of b-diketonates of some metals with temperature: 1 - Cr(AA)$_3$; 2 - Fe(AA)$_3$; 3 - CR(TFA)$_3$; 4 - Fe (TFA)$_3$; 5 - Cr (GFA)$_3$; 6 - Fe (GFA)$_3$; 7 - Mo(GFA)$_3$; 8 - W(GFA)$_3$; 9 - Ti(AA)$_4$; 10 - Zr(AA)$_4$; 11 - Hf(AA)$_4$; 12 - Hf(GFA)$_4$; 13 - Hf(FOD)$_4$; 14 - Hf(GFEN)$_4$; 15 - Ti(GFEN)$_4$.

On the basis of the experimental data, the p-T diagrams have been constructed and the parameters of the vapor-formation processes have been calculated for the series of b-diketonates of p- and d-metals and for some adducts of b-diketonates of Mn (II), Co (II), Ni (II) and Cu (II). In Figure 14 and Table 20 the data are presented for some b-diketonates of metals that were not studied earlier.

The fact that the obtained experimental data and the values of H_T^0 and S_T^0 calculated on their basis may be classified as the heterogeneous process ML_m (sol., liq.) $\rightleftarrows ML_m$ (gas) is confirmed by the calculations of the molecular masses of some b-diketonates and their adducts in the gas phase for the given p-V-T for the region of unsaturated vapor. These values are in good agreement with the theoretical values of the molecular masses of b-diketonates and confirm transition into the vapor phase of monomeric chelates.

The data analysis shows that the obtained laws governing the transition of the studied b-diketonates of p- and d-metals are in good agreement with the conclusions made in earlier publications. The introduction of fluorinated a-substituents increases the volatility of all the complexes, while decreasing their temperature interval of thermal stability.

Table 20. Thermodynamic parameters of transition of
some b-diketonates of m metals in vapor phase

Complex	Process	Temperature interval °C	Coefficients of equation lg p=B - A/T		$H°_T$, kJ/mole	$S°_T$, J/mole◊deg
			A	B		
Ti (AA)$_4$	sbl	170-190	66201±165	13.4±0.8	118.7±3.2	202.2±15.3
Ti (TF H)$_4$	vpr	80-115	4199±115	12.7±0.7	80.4±2.2	188.1±13.4
Zr(AA)$_4$	sbl	150-200	5999±117	13.1±0.9	114.8±2.3	195.6±17.2
Hf(AA)$_4$	sbl	150-210	6697±130	14.8±1.1	128.2±2.4	229.0±21.0
Hf(FFA)$_4$	vpr	60-110	6025±140	18.3±1.4	115.3±2.6	296.1±26.7
Hf(FOD)$_4$	vpr	60-110	4643±163	14.4±0.8	88.9±3.2	220.0±15.3
Hf(GFZN)$_4$	vpr	75-115	4099±172	12.5±0.6	78.5±3.3	184.5±11.5
Cu(Gep D)$_4$	vpr	150-215	5801±107	13.1±0.9	111.1±2.1	196.3±17.2
Cu(Nop ?)	vpr	135-220	5701±116	12.8±0.7	109.1±2.3	190.1±13.4
Cu (TR D)$_2$	vpr	120-210	5301±108	11.7±0.6	101.5±2.2	169.1±12.1
Mn (GFA)$_2$◊2DMFA	sbl	100-210	2199±96	5.8±0.9	42.0±2.0	57.3±17.3
Co(FFP)$_2$◊2DMFA	sb	110-215	2525±125	5.9±0.2	42.3±2.4	58.2±5.7
Ni(GFA)$_2$◊2DMFA	sbl	105-200	2449±102	7.1±0.4	46.9±2.0	79.9±6.3
Mn(GFA)$_2$◊2TBFO	sbl	105-205	2101±114	6.7±0.3	40.2±2.2	73.7±7.1
Co(GFA)$_2$◊2TBFO	sbl	105-205	1974±116	5.5±0.3	37.8±2.4	49.4±6.2
Ni(GFA)$_2$◊2TBFO	sbl	105-205	1925±100	5.3±0.2	36.8±1.9	45.6±5.8

*-CnH_{2n+1}

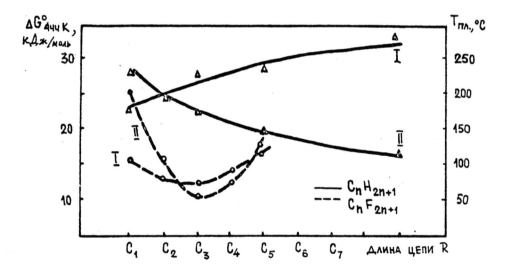

Figure 15. Effect of length of chain of aliphetic substitute on volatility (I) and melting point (II) of bis-b-diketonates of copper (II).

The authors have noted that for the b-diketonates with the straight chain of one of the a-substitutes ($-C_nH_{2n+1}$) elongation of this chain with concurrent reduction of the melting point of the complex results in a decrease in its volatility, whereas the introduction of the similar fluorinated substituents causes the volatility curve to pass through the maximum (Figure 15).

Formation of adducts somewhat decreases volatility, but considerably stabilizes the complex extending the region of its thermal stability (for perfluorinated b-diketonates).

It has been shown that the major factor affecting volatility of the b-diketonates is the nature of the ligand, and in particular the sterie and inductive effect of its a-substitutes and the nature of the donor atoms of the coordination site of the chelate.

In terms of the capability to effect the reduction of volatility of the complex the ligands and the donor atoms in the coordination node of the complex can be arranged into the following series:

GAF > DFG > GFEN > TFA > AA > DPM;

[0,0] > [0,N] > [0,S].

The effect of the nature of the metal is less pronounced and the general correlation in terms of volatility is more difficult to conduct, though it is seen that the following relationship takes place: $M(b-dik)_4 >> M(b-dik)_3 > M(b-dik)_2 \lozenge 2D > M(b-dik)_2$.

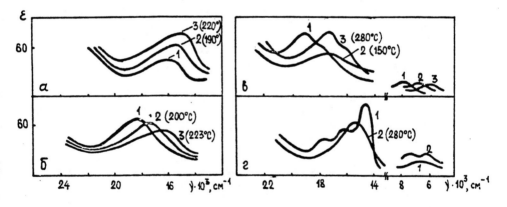

Figure 16. Effect of nature of metal and substitute of b-diketone on decomposition temperature of b-diketonate complexes in gas:
a) $M(TFA)_3$, M-V, Cr, Mn, Fe, Co;
b) $M(AA)_3$, M-Al, Ga, In;
c) $M(GFA)_4$, M-Ti, Zr, Hf;
d) $Al(B-dik)_3$ and $Cu(B-dik)_2$

For an analysis of the stability of the b-diketonates of metals with variations of the decomposition temperature the obtained data (Figure 16 a, b, c) were compared for the b-diketonates of the same structure and composition. It is seen that the values of T_{decomp} in the gas volume correlating with the average energies of dissociation of metal-ligand (B) for b-diketonates of the p-metals decrease with increasing atomic number of the element of the same group (Al>Co>In), whereas for the transition metals the reverse situation is observed (Ti<Zr<Hf). In the case of transition along the 3d-series of elements the temperatures of decomposition (and correspondingly the D bonds) decrease with increasing atomic number.

The introduction of various a-substituents in the b-diketone, i.e. reconstruction of the conjugated electronic system of the complexes also changes the temperature of decomposition of complexes in the volume.

The Tdecomp decreases as a linear function of the electron-donor force of the a-substitutes of the b-diketone of the complex.

The obtained data show the presence of the symbate changes in thermodynamic parameters of the vapor-formation processes and processes of decomposition in a volume of b-diketonates of metals and in characteristics of the electronic and geometrical structure of their individual molecules.

c) Basic criteria of volatility and thermal stability of b-diketonates of metals [38]

Volatility and thermal stability of the b-diketonates is ensured by combining their specific electronic and geometrical structure.

As the authors have already noted, the specific feature of the electronic structure of b-diketonates is the presence of various sub-systems of bonds in the complex: the stable cyclic delocalized s- and p-system of the MO of the complex and two dynamic s- and p-systems of the HOMO of the complex. These sub-systems partially interacting with one another are responsible for ensuring the existence of the b-diketonate as a stable compound and for forming its specific properties (volatility and stability).

The specific features of the geometrical structure of the b-diketonate complexes are: the presence of the relatively rigid flat cyclic skeleton of the ligand with the concurrent "dynamic non-rigidity" of the coordination site which manifests itself in the ability to realize the intramolecular transitions of the central atom of the metal; the presence of the complex periphery formed by the a-substituents of b-diketone which screens the coordination node and forms the "island"-like structure of complexes; possibility of opening the ring, i.e., under certain conditions possibility of transition from bidentate coordination of b-diketone into the monodentate one.

None of the above-mentioned specific features of the electronic or geometrical structure of b-diketonates is critical, the specific properties inherent to the b-diketonate complexes of metals or to their N- or S-analogs are formed only by their combinations and interaction.

The basic criteria which make the b-diketonates capable of going easily into the vapor phase at relatively low temperatures without destruction of the molecular structure (vapor pressures ranging from 10 to 1000 torrs at 50 to

250°C) and existing with a high stability in the vapor phase within the relatively high temperature interval are as follows:

- the presence of a high degree of covalence of the metal-ligand bond, (the ratio of the covalent to the conic component of the bond central atom-ligand>1), ensuring delocalization of the electron density throughout the entire complex and the absence of the uncompensated Coulomb charge on the central atom requiring the stable existence of the complex the latter's neutralization by the additional ligand;
- the presence of the dynamic conjugated-bond system ensuring dissipation of a considerable amount of thermal excitation energy in the atoms and bonds of the complex molecule as a hole;
- the presence of the molecular periphery of the complex consisting of the end groups of atoms, ligand's substituents possessing relatively free torsional rotation and weakly interacting with one another and with other similar groups (for instance, the fluoro-substituted a-substitutes);
- the possibility of screening the coordination center, (both the metal and the donor atoms), by the group of the peripheral atoms. Within the framework of the proposed criteria the problem of volatility and stability of complexes can be solved and the complexes with desired properties can be formed;
- introduction of the fluorinated a-substitutes into the b-diketone makes it possible to reduce the vaporization temperature and the temperature at which the thermal decomposition of the complex begins;
- introduction of branched-bond alkyl substitutes into the b-diketones makes it possible to increase the volatility of coordination unsaturated complexes and, in general, to increase the interval of thermal stability of all complexes in the gas phase;
- introduction of additional donor ligands into the coordination unsaturated complexes makes it possible to avoid their oligomerization, reduce the vaporization temperature and increase the interval of their thermal stability in the gas phase;
- the joint introduction of the electron-acceptor substitute and formation of adducts makes it possible, for instance, to direct the thermal decomposition of b-diketonates along the way of predominant evolvement of the metal phase without destruction of the ligand.

3. GAS-TRANSPORT, HOMOPHASE, HETEROPHASE PROCESSES WITH PARTICIPATION OF HIGH-VOLATILE B-DIKETONATE COMPLEXES [38]

The brief objective of this section is to substantiate the possible way of utilizing the volatile b-diketonates of metals. In the last paragraph of the preceeding section the fundamental possibility of purposeful thermal decomposition for evolvement of metal without destruction of the ligand was mentioned. The processes of dissociation of b-diketonate complexes can follow this course, (nomolytic dissociation), with formation of neutral particles of metals and radicals of ligands and the course of heterolytic dissociation with the formation of the molecules of oxygen- and carbon-containing compounds of metals and the products of thermal fragmentation of ligands.

The authors have conducted mass-spectroscopic studies, (electron impact and chemical ionization), of the effect of the nature of the a-substitute of b-diketone and the nature of the central atom (Al, Ga, In, Ti (III), (IV), Zr (IV), Hf (IV), Cr and Fe (III), Co (II), Cu (II) and Zn (II)) on the paths of fragmentation of their b-diketonates.

In all cases, the initial stage of fragmentation is a detachment of one of the ligands, following the process of decomposition of the formed fragment follows two parallel courses: the first one leads to the formation of the individual metal, the second one through decomposition of the ligands leads to the formation of oxygen-carbon-containing compounds of the central atom, with the direction of the decomposition course being dependent on the nature of the metal and the character of the a-substitutes in b-diketone.

For trans b-diketonates of Al (III), Ga (III) and In (III) the tendency to form an individual ion of metal becomes more pronounced when proceeding from aluminum to indium.

For the investigated 3d-metals, obtaining ionized metals is characterized by rich and diverse fragmentation, with the general pattern followed by the fragmentation process being formed by the ions containing the M-F bond.

The results of studies of the processes of some b-diketonates of metals by the differential-thermal analysis method in the air of an argon atmosphere have been compared with the results of the mass-spectroscopic and gas-chromatographic analysis of the products of their thermal decomposition. The tendency has been observed towards a decrease in the temperatures of thermolysis of the fluorinated b-diketonates of both the p- and d-metals. Formation of the adducts sharply increases the thermal stability and volatility of such complexes.

The analysis of the thermal decomposition products shows that their composition, (acetone, CO, CO_2, methane, ethane, butene, etc.), is basically similar for most of the complexes, with the composition of the thermolysis products for the 2.4-pentandiades of metals being richer than that for their fluorinated derivatives. This is caused by the lower decomposition temperatures of fluorinated b-diketonates and, as a consequence, by a higher yield of the undestroyed ligand.

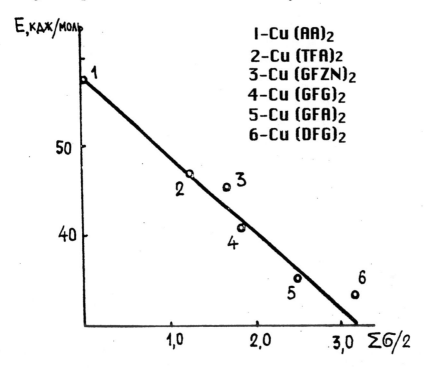

Figure 17. Relationship between energy of activation of thermal decomposition of b-diketonates of copper (II) and the value of inductive constant of substitutes in ligand; 1 - Cu(AA)$_2$; 2 - Cu (TFA)$_2$; 3 - Cu(GFEN)$_2$; 4 - Cu(GFG)$_2$; 5 - Cu(GFA)$_2$; 6 - Cu(DFG)$_2$.

The 2.4-pentandionates of metals with their high decomposition temperatures, (>300°C), are characterized with a deeper decomposition depending on the nature of the central atom, its crystal activity, its ability to form oxides, carbides, etc.

The authors have studied the kinetics of the process of thermal decomposition of b-diketonates in the closed and open systems and have calculated the rate constants of decomposition and have observed that the introduction of a fluorinated a-substitute increases the rate of the reaction of thermal decomposition of b-diketonate.

The energies of activation of the thermal decomposition reaction are close in magnitude for similar-type complexes irrelevant of the nature of the central atom. They decrease rapidly when proceeding from the coordination saturated to coordination unsaturated complexes, which is caused by a weaker screening of, and easier access to, the coordination site of the latter complexes during formation of the solid phase.

The obtained kinetic parameters of thermal dissociation for the bis-b-diketonate complexes of copper (II): Cu (AA)$_2$, Cu(TFA)$_2$, Cu(GFA)$_2$, Cu(GFG)$_2$, Cu(DFG)$_2$ and Cu(GFEN)$_2$ have shown that the energy characteristics of the process vary as linear functions of the electron state of the molecule of the complex.

The effect of the a-substitutes on the energy of activation (Figure 17) can be shown as a series: DFG<GFA<GFG<GFEN<TFA<AA.

The obtained curves showing the maximum rates of growth of the aluminum oxide film at preset temperatures, (heterolytic dissociation), have made it possible to disclose the basic kinetic regions of such thermal decomposition and to determine the limiting phases of this process.

The basic kinetic regions are found to be:

- the region with a high activation energy, (about 95 kJ/mol), where the limiting processes are the formation of aluminum oxide in the adsorbate and its transition into the growing film. This region is characterized by low rates of growth of the film and its amorphous structure;
- the region where the limiting process is the activated adsorption. Because of the chain character of the decomposition and crystallization reactions, this region exhibits relatively low values of activation energy, (about 30 kJ/mol), and high rates of growth of the grain-oriented film with good adhesion to the substrate;
- the region with the limiting process of adsorption of the molecules Al_2O_3 from the volume with the activation energy close to zero and, as a consequence, formation of a loose film with poor adhesion to the substrate.

From a technological viewpoint the most promising turns out to be the kinetic region where the limiting process is the activated adsorption of the complex on the surface of the substrate, in which case the energy of activation of the thermal decomposition process is relatively low and the rate of the growth of the films with good adhesion is considerably high.

The purposeful formation of desired properties of highly volatile b-diketonates of metals makes it possible to define some files of their utilization for obtaining materials and coatings by the chemical vapor deposition (CVD) method.

Examples are the methods for obtaining high-dispersity powders of copper and nickel from their 2.4-pentandionates and 2.4-iminopentanates in the multi-atom alcohol medium; copper, nickel and cobalt coatings from the adducts of their perfluorinated b-diketonate complexes with glycol or nitrogen-containing bases, (without additional reduction by hydrogen); optically- and acoustically-active zinc oxide, dielectric aluminum oxide films and transparent electro-conductive indium oxide films by low-temperature plasma-chemical decomposition of their 2.4-pentandionates and trifluoro-2.4-pentandionates; ultra-high-dispersity extra-pure sized-grain powders of oxides of aluminum and gallium by gas-flame burning of their b-diketonate complexes; flame- and erosion-resistant aluminum- and hafnium-oxide coatings; shielding gas media for heat-treatment of iron-carbon alloys, etc.

V. COORDINATION COMPOUNDS IN LASER CHEMISTRY [58-72]

The laser chemistry is essentially a branch of the high-temperature chemistry, since using the resonant techniques of laser excitation of desired energy levels in atoms and molecules it can, at equilibrium room temperatures, induce interesting high-temperature physical and chemical processes due to a high non-equilibrium thermal and selective oscillatory excitation [1-6].

Whereas at initial stage of practical use of lasers they were regarded as only sources of high-power light energy, presently there is a tendency towards differentiation in utilization of their unique characteristics which, apart from power, include monochromaticity, coherence and sometimes pulse shaping frequencies [58].

But even the seemingly simple laser-thermal, (thermochemical) processes induced by radiation power are rather singular in character: the volume can only be heated where the reaction takes place, thus avoiding undesired heating of all the parts of the equipment and consequently avoiding a more active interaction of reagents with heated walls, processes in a non-isothermal regime can be conducted, etc. However, in principle, lasers as sources of thermal energy can be rationally used only in the cases of need for exotic processes or for obtaining extra valuable substances and materials.

At the same time, the two other unique characteristics of laser radiation, namely high monochromaticity and coherence open the opportunities for fundamentally conducting new physical and chemical processes allowing considerable amounts of energy and resources to be saved. The fact is that the high monochromaticity allows for the conducting of non-equilibrium photo-chemical processes with the molecules of a strictly definite type subjected to selective excitation by resonant radiation of lasers rather than with the entire reaction medium, (intermolecular selectivity), and sometimes only with selectively-excited oscillatory modes of these selected molecules (intra-molecular selectivity). So, these processes are expected to be energy- and cost-saving.

In its turn, the coherence of laser radiation allows for the large-scale introduction of new technologies in such fields of science and technology as recording of information, producing of optical elements, microelectronics, computer engineering, etc.

The authors will try to illustrate the afore-mentioned with examples of the researches conducted in our laboratory* in the field of the laser chemistry of coordination and high-coordination compounds in the gas phase, in melts and in polymerizing and solid phases [58].

---

* The authors acknowledge the participation of A.V. Boltsman, A.F. Gurko, B.I. Lutoshkin and others in the research.

1. LASER CHEMISTRY OF HIGH-COORDINATION COMPOUNDS IN GAS PHASE

a) According to the modern concepts, classified as the thermal laser-chemical reactions should mainly be the reactions of the molecules absorbing the radiation of the low- and medium-power (up to about 100W) continuous-wave infrared lasers. Naturally, it is expedient that thermal reactions be realized with the help of thermal "oscillatory" lasers with a minimum power. Since in this case the high-precision resonance of laser radiation with the oscillation frequency of the excited molecules is not required, the number of suitable compounds turns out to be large enough.

Among them; the most intensively studied is the interaction of SF_6, BCl_3, SiH_4, etc. with oxygen, hydrogen and saturated and unsaturated hydrocarbons. The kinetic of these reactions was found to differ considerably from the conventional thermal reactions by the strong effect of the frequency of laser radiation on the yield of the reaction products, by the presence of the reaction threshold in terms of the intensity of laser radiation and by other factors. The unconventional behavior of laser-chemical reactions is caused by the establishment of feedback between chemical and thermal degrees of freedom due to variations in bond absorption in the course of chemical reaction.

It is the presence of this feedback that is responsible for the complex dynamic character of such synergetic laser-chemical reactions: their instability, thresholds of reaction rates, self-excited oscillations, etc.

These and other questions have been discussed, for instance, in studying the laser-chemical reactions with participation of BCl_3 [59-61]: $BCl_3 + H_2 \rightleftharpoons BCl_2H + HCl$.

It should be noted that dichlormonoborane (BCl_2H) fails to be synthesized from BCl_3 and H_2 (or CH_4) in a high-frequency plasmatron and in electrical discharge, whereas under the action of an IR-laser of about 30W in power its yield reaches 50%. Under the action of the IR-laser, (200-500 W/cm^2 in intensity), the reaction $BCl_3 + SiF_4 \rightleftharpoons SiF_3Cl + BCl_2F$ and other reactions have been recognized.

Of particular interest is the sensibilized thermal laser synthesis, since it allows the range of possible reactions to be considerably expanded. Used by us as the IR-sensibilizers were the multiatom and coordination molecules intensively absorbing the IR-band laser radiation and resistant to the thermal effect of this radiation, for instance SF_6 [62] and others. The synthesis of formaldehyde by soft oxidation of methane by the oxygen in the presence of a sensibilizer [58] may serve as an example of such laser process.

The afore-mentioned examples of characterized mainly thermal laser-chemical reactions.

b) The selective photosubstitution of ligands in the coordination sphere of complex-forming metal [58, 63] may serve as an example of the reaction of the coordination compounds utilizing the *monochromaticity* of laser radiation:

$$ML_3 \xrightarrow{h\nu} (ML_3)^* \rightarrow ML_2 + L,$$
$$ML_2 + X \rightarrow ML_2X,$$

where M is the metal, (lantanoid or actinoid), L is the chelatetype ligand, (B-diketone, ethylenediamine, or the like), X is the new ligand being introduced.

The new product, ML_2X, which has been formed as a result of the selective photo-substitution under the effect of the strictly-resonant laser radiation differs substantially from the original reagent ML_3, hence it can serve as a basis for the separation of metals similar in their properties.

Regarding the realization of the intramolecular selectivity processes based on absorption of monochromatic IR-laser radiation by the oscillatory modes of molecules resonant in frequency, they are observed in the cases where the molecule is exited by high-power, (about MW/cm^2, GW/cm^2), pulse lasers during the time intervals substantially shorter than the time of its relaxation.

The difficulty in heating of the selected chemical bond, and consequently its dissociation, by laser radiation is accounted for by the statistical monomolecular decomposition theory (RRKM). The molecule dissociates along the weakest chemical bond irrespective of the oscillatory mode through which the laser energy is applied to it, which is cased by the interim-mode energy randomizing process for a period of about 10^{-10}s [58, 64].

c) Growing interest has lately been focused on development of the lasers using the vapors of complex organic, metal-organic and inorganic compounds. Compared to solid-state lasers these lasers should possess a number of undisputable advantages: homogeinity of the medium, possibility of its pumping through the active volume, and consequently obtaining of high power of generation. The publications are known to be dealing with the research into the possibility of developing lasers using the ions of rare-earth elements contained in the compound with halogens. The interest shown in these compounds where rare-earth elements are in the stage of oxidation (III) is accounted for by the fact that with their participation the four-level laser version of active medium can be realized.

The publication [65] is devoted to studying the problem of deactivation of the state $^4F_3/2$ of the ion of neodimium (III) incorporated in the complex $NdCl_3$ ($AlCl_3$) in the vapor state. (It should be noted that evaporation of the compounds of this type requires the temperature of a few hundreds of centigrades, which considerably complicates both the experimental technique and development of laser).

Publication [66] studied the fluorescence of vapors of a complex of neodimium and terbium with dipivaloilmethane (NdL_3 and Tbl_3) and the kinetics of fluorescence at a transition of $^4F_3/2 \AE ^4Il_{1/2}$ (1.06 mm) of chelate NdL_3 in vapors (excitation was performed by the pulse laser using the solutions of organic dyers in the intensive absorption band in the neighborhood of 5872Å; at temperatures of 150 to 230°C concentration of vapors of NdL_3 was about (0.002 to 2)$\lozenge 10^{17}$ cm^{-3}). Publication [67] deals with the problem of laser-induced fluorescence of vapors of complexes ErCl$_3$ (ACl$_3$)X (A=Al, Ga, In with x=1 to 4) at high temperatures. In all these publication devoted to studying fluorescence the vapors of the compounds were excited by coherent radiation.

The authors have studied the fluorescence of vapors of b-diketonate complex of euorpium excited by an noncoherent radiation source. Taken as an object being studied was europium dipivaloilmethanate EuL$_3$ (L is residue of type (a); coordinate site of the unhydrous complex EuL$_3$ in vapor has the form (b) [68]:

The temperature of the cuvette was maintained at about 250 °C, i.e. it was lower than the temperature of decomposition of the complex. The monochromator was adjusted to the wavelength of 6120Å corresponding to the transition $^5D_0 \mathcal{E}^7F_2$ of the ion of europium in the complex.

Comparison of the oscillograms has shown that in the presence of vapors of EuL$_3$ in the cuvette, the signal is stronger by about 30% than in the absence of the vapors. This is accounted for by fluorescence of Eu (III) from the level $^5D_0$. In the case of departure by ±200Å from the line 6120Å of the monochromator the intensity in the cuvette filled with vapors of the complex falls by 30%.

If the dielectric mirror with the transmission coefficient of about 35% is placed between the cuvette and the inlet slot of the monochromator at wavelength of 6120Å, the intensity of the total background and luminescence signal decreases by 70%. If the second dielectric mirror with the reflection coefficient of about 65% is places at the opposite end of the cuvette (creation of the laser resonator), the intensity of the signal increases about 3 times, i.e. reaches the previous magnitude. Thus, at concentration of vapors of EuL$_3$ equal to about $3 \lozenge 10^{17}$ cm^{-3} the presence of radiation feedback results in a growing intensity of the fluorescence signal. The obtained results show that by increasing the pressure of the vapors in the cuvette to a certain value it is possible, in principle, to attain generation at the transition $^5D_0 \mathcal{E}^7F_2$ of the ion of europium (III) contained in the studied complex.

2. LASER CHEMISTRY OF COORDINATION COMPOUNDS IN MOLTEN SALTS

As an example of such pioneering research the authors will present the study of temperature- and thermal-conductivity of melts with coordination compounds using the unique characteristic of laser radiation-coherence [69]. It has already been shown that thermal processes are the main processes responsible for formation and decomposition of the dynamic diffraction grating in liquids, in particular in dyers. In addition, the time characteristics of the dynamic holograms have been studied in detail and the experiments have been conducted for checking the possibility of redistribution of intensities of power of the beams forming the dynamic holograms and finding of optimal conditions for attaining the maximum diffraction efficiency.

On the basis of the known values of temperature and thermal conductivity a good agreement has been reached between the theoretical and experimental results. The inverse problem-obtaining of thermal physical parameters from the results of studies of relaxation characteristics of gratings in melts was studied by the authors for the first time. The aim of these studies is the decide whether or not the dynamic holography method can be applied to studying temperature and thermal conductivity of liquids, including high-temperature melts of aggressive salts.

This method of studying thermal conductivity should be regarded as a nonstationary method with a pulse-type source of heat and periodic variation of temperature over the specimen being studied in conformity with the formed diffraction grating. Non-stationary methods are known to be characterized by higher accuracy and require less time as compared with stationary ones. In this case the proper instrument time of measurement is as small as a few milliseconds.

When comparing the holographic technique with conventional optical ones, in particular with the interferometrical methods of studying thermal-physics parameters, it is worth noting that it is distinguished by mathematical simplicity: most of the calculations on transforming the field of the refractive index into the temperature field are performed "automatically" on restoring the diffraction grating by the test radiation. If the boundary conditions are neglected, the time characteristics of the grating can be described by the following equation:

$$\frac{1}{\alpha} \cdot \frac{\partial T(X_1 t)}{\partial t} - \frac{\partial^2 T(X_1 t)}{\partial X^2} = \frac{I(X_1 t)}{Kh}$$

where $T(X,t)$ is the temperature distribution, $a = K/Cp$ is the temperature conductivity coefficient; K is the thermal conductivity; C is the heat capacity; p is the density; h is the thickness of layer of the medium;

At equal amplitudes of the waves forming the grating, $I(X,t)$ takes the form

$$I(x,t) = I_0(t)[1 + \cos(2\pi x / \lambda)],$$

where $L = 1/2Sinq$ is the period of distribution of the interference maxima; is the wavelength of light; 20 is the angle between the beams.

If L and h are small as compared with the longitudinal dimensions of the grating the solution of the equation to the accuracy of the constant takes the form

$$T(x,t) = \frac{I_0(t)}{c\rho h} + \frac{I_0 \Lambda^2}{4\pi kh}(1-e^{-t/\tau})\cos\frac{2\pi x}{\Lambda}$$

where $t = L^2/4p^2$ is the time constant of thermal diffusion. From the analysis of the equation it follows that on restoring the formed diffraction grating by the test radiation, the relaxation of the intensity of the first diffraction order referenced to the unit falling flux can be written as

$$\eta = J_1^2(\varphi_0 e^{-t/\tau}),$$

where h is the diffraction efficiency; J_1 is the Bessel's function of the first order, F_0 is the value of the phase run-on in the initial moment of time. Having obtained the experimental relationship, the authors can easily calculate from the equation for (t) and, consequently calculate the thermal conductivity at the known heat capacity and density.

The measurements were first conducted for liquids, (water, ethyl alcohol, acetone), with well-known thermal-physics parameters. Table 21 shows the data on the mechanism of absorption of laser radiation. Since the liquids being studied have no absorption in the neighborhood of 1.06 mm the tinting additives $Cu(NO_3)_2$ or $CoCl_2$ were introduced. The comparison of the obtained results with the known data points to the discrepancy of about 30% which is accounted for by both the inaccuracies inherent to all the techniques and by their fundamental differences. Following this, studies were conducted for the complex ions in melts (Table 21).

Table 21. Thermal and temperature conductivity of investigated solutions and melts

Substance	Absorption mechanism	Thermal conductivity x10^{-4} J/cm◊s◊K		Temperature conductivity x10^{-3}, cm^2/s
		Experimental results	Table data	
C_2H_5OH(5% $Cu(NO_3)_2$, 20°C)	single - background	12.5	18.5	0.64
Acetone (5% $Cu(NO_3)_2$, 20°C)	single- background	11.7	17.2	0.68
H_2O(10% $Cu(NO_3)_2$, 20°C)	single- background	53.8	56.7	1.28
$Co(NO_3)^{2-}_4$in K, Na/NO$_3$, entectics (250°C)	single- background	–	–	1.29
$K_2Cr_2O_7$, (400°C)	double- background	–	–	1.20
$CoCl^{2-}_4$ in Li, K/Cl, entectics	single- background	–	–	2.42
400°C		–	–	2.15
415°C		–	–	2.07
470°C		–	–	1.98

In conclusion, the authors note the principal advantages of the method: rapid (about 10^{-3} s) determination of thermal-physics parameters, the possibility of remote measurements, absence of contact between the equipment and high-temperature aggressive melts, etc.

The necessity for tinting the liquids for absorption of laser radiation arises only due to the unavailability of the proper lasers, for instance those radiating in the infrared band and shall be eliminated as soon as such lasers become available.

3. Laser Chemistry Of Coordination Compounds In Polymerizing And Solid Phases [70-73]

This section illustrates utilization of the coherence of laser radiation in polymerizing and solid systems containing coordination compounds of metals. The concept of coherence has become widely used in science and technology with the advent of high-power sources of coherent radiation-lasers. In practice, coherence is obtained by proper positioning of two laser beams radiated from a single source. It manifests itself as a clearcut interference pattern with the sinusoidal distribution of illuminance in the form of light and dark lines and is depicted as the holograms recorded by various light-sensitive materials- halogen-silver, photochromic, ferroelectric, optomagnetic, thermoplastic, photo-semiconductive, etc.

The cross-linked gelatine ensuring high resolution power and diffraction efficiency has turned out to be one of the most effective material for recording holographic information. Equally promising are the photopolymerizing layers combining high sensitivity and diffraction efficiency with the simplicity of preparing the photosensitive layers on their basis.

Whereas in the former and latter materials, (particularly for bichromated gelatine), there is sufficiently ample information, no publications are yet known concerning the material combining the properties for both of them.

It is very tempting from both scientific and technical viewpoints to create and study such systems for the holographic recording of information which would combine the advantages of the photo-laser-polymerizing systems, (acrylic acid compounds are considered as such in the present publication), with the specific positive properties of such natural polymer matrix as gelatine.

The authors have selected for study: photographic gelatine as a medium and the component to be cross-linked; acrylic acid compounds, (amides and salts of calcium, cobalt, nickel, etc.) as cross-linking complex-forming and at the same time photopolymerizing reagents together with sensibilizer-methylene blue and the electron donor-triethyl amine.

When the gelatine film matrix is treated with aqueous solutions of amides of acrylic acid, (acrylamide - AD and methylene-bis-acrylade-MBAD), the latter are absorbed by gelatine. These processes manifest themselves more clearly when gelatine is replaced with the specific aminoacidoxyproline and when the experiments are conducted not only in the film but in the liquid-phase volume as well.

The IR-band frequencies are noted which disappear or decrease in intensity as a result of laser radiation of the components of the reaction: $CH_2=CH$ groups of amides and OH- and NH_2-groups of oxyproline and gelatine, as well as the frequencies of the C-O groups of common either appearing as a result of the interaction of OH-groups of gelatine with the vinyl group of MBAD.

Thus, the laser radiation first causes the opening of the double bonds of the vinyl groups in the molecules of amides with the subsequent cross-linking specific for each amide:

- In the gelatine system - AD the oligomer-dimer is first formed due to the interaction of the amino-group with the vinyl group of other molecules of AD which is then attached to the macro-molecules of gelatine along its amino- and hydroxyl-groups by its vinyl groups thus forming the "comb-shaped" structure.
- In the gelatine system - MBAD cross-linking under the effect of the laser is essentially the linear, (partially three-dimensional), cross-bonding due to the interaction of the opened vinyl groups of the longer molecule of MBAD with amino- and hydroxyl groups of the gelatine with resultant formation of cross-linked planar and three-dimensional structure.

The studies of the kinetics of the swelling of the original gelatine and that which was cross-bonded under the laser effect with AD and MBAD confirms the data of IR-spectroscopy: the degree of swelling (Q_{max}) decreases when proceeding from gelatine to the gelatine-AD system, gelatine-MBAD due to the fact that it is more difficult for water to penetrate into the three-dimensional structure than into the "comb-shaped" one.

Table 22. Macromolecular parameters of photolayers of gelatine
layer-cross bonded with amines of acrylic acid

Cross bonding polymerizing reagent	Parameters of cross-bonded compound				
	$p, g/cm^3$	Q_{max}	K_{swell}, S^{-1}	$dQ/dt, S^{-1}$	$M_{Cl}mol$
AD:$CH_2=CH-$ C	1.34	6.91	$6.90 \cdot 10^{-5}$	$3.26 \cdot 10^{-4}$	51400
MBAD: $(CH_2=CH-$	1.62	2.42	$4.00 \cdot 10^{-5}$	$8.46 \cdot 10^{-5}$	7300

The afore-mentioned qualitative specific features of the structures have been quantitatively confirmed by the studies and calculations of the macro-molecular parameters of these media (Table 22): their density (p), degree of swelling (Q_{max}), rate of swelling (dQ/d), molecular mass of the fragment with the adjacent "dopes" and "cross-links" (M_c) of the molecule.

The deeper the cross-linking processes in the system, the greater the density by which they are characterized, the less the swelling rate and the less the value of the molecular mass of the fragments between the cross-links, and hence these system should be *a priory* distinguished by better photooptical parameters.

The studies of the sensitometric characteristics of the photo layers of gelatine laser-cross-linked with the help of AD and MBAD (Table 23) has shown that this really is the case and that in the latter case even a more light-sensitive system (30 mJ/cm^2 instead of 100 mJ/cm^2) with higher resolution power (l/d) can be obtained.

Thus, the established relationships between the sensitometrical and macro-molecular parameters of the photolayers cross-linked by chemical laser-assisted cross-bonding depends on the frequency of cross-linking of the sections of photosystems, i.e. on the values of M_c. It depends consequently on the number of vinyl groups in the molecule of the photo-crossbonding reagent, and in the general case, on the specific number of the functional groups taking part in the laser-assisted cross-bonding.

From this, one should try to use as the reagents for cross-bonding gelatine with the help of lasers the molecules with great number of, for instance, vinyl groups.

However, since propylene-bisacrylamide's solubility in water decreases in the series of ethylene, and the photocomponents, (from the conditions of obtaining photo media), should be sufficiently soluble in water, the most promising way appears to be the utilization of the salts of metals or complexes of metals whose anions or ligands contain more vinyl groups as a whole, rather than increasing the number of vinyl groups in the extended chain of organic compounds.

Table 23. Optical characteristics of photolayers of photolayers of gelatine laser-cross-bonded with amides of acrylic acid

Cross-bonding polymerizing reagent	Parameters of cross-bonded compound h, %	E, mJ/cm^2	S, m^{2}J	l/d, mm^{-1}
AD:CH$_2$=CH-C	33	100	$5.9\lozenge10^{-3}$	1580
MBAD:(CH$_2$= CH-C	33	30	$20\lozenge10^{-3}$	1210

As an example let us consider the processes occurring under the effect of laser radiation of the gelatine matrix with calcium and copper (II) acrylates.

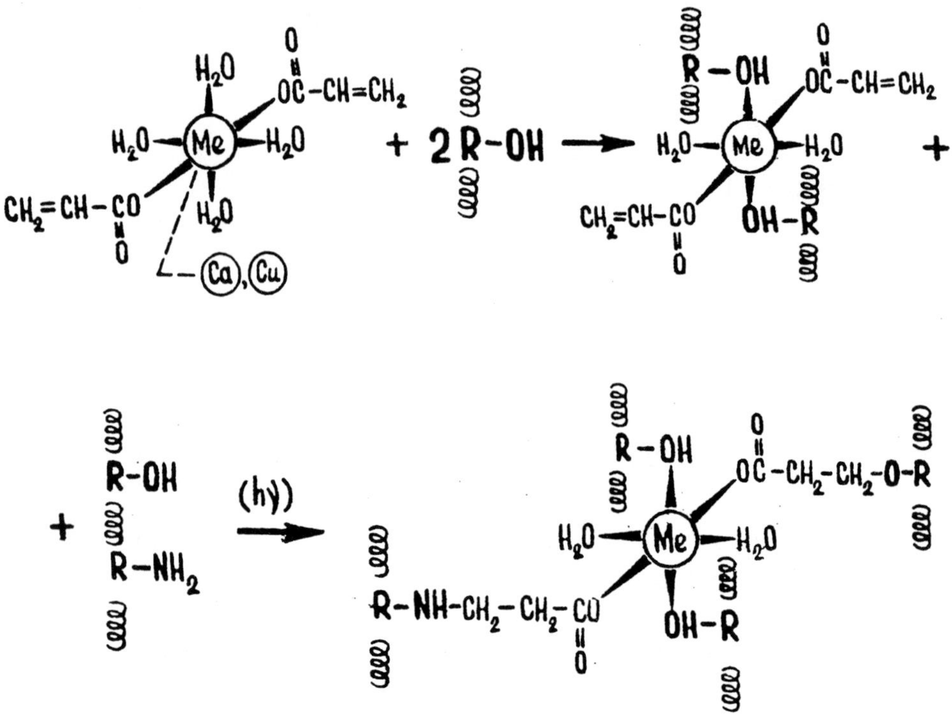

Figure 18. Complex formation and cross-linking of gelatine with salts of acrylic acid and radiation of He-Ne laser.

With the use of the IR and electronic absorption spectroscopy, (for copper), i has been established that the acrylates of metals are absorbed from aqueous solutions by the gelatine matrix and that the coordination site Me^{2+} is first formed from acrylate-ions, molecules of water and two hydroxyl groups of oxyproline of gelatine (shift of IR-bands of absorption of oscillations of OH-groups of oxyproline). Under the effect of laser radiation the double bonds of the vinyl group of acrylate ions become opened in the formed and already partially cross-bonded macrocomplex Me^{2+} and these groups interact with another two OH- and H_2-groups of oxyproline of gelatine (Figure 18). It was noted that the intensity of the IR-strips of the bonds of $CH_2=CH$ - groups and CH; NH_2-groups has decreased and that the oscillations of the C-O bond of the ether group have appeared as a result of the interaction of the vinyl group in the complex with the OH-group of oxypynoline. As a result of such double cross-bonding due to complex-formation by metal and laser-chemical cross-bonding photo layers with an even higher density are formed (Table 24) and with smaller or commensurable with MBAD values of Q_{max} and M_c. The double cross-bonding results in the obtaining of the optimum photo layers with the participation of calcium acrylate requires tenths of mg (~0.6 mg), whereas the same result in the case of amides of acrylic acid can be obtained with a few tens of mg (75 mg AD).

Table 24. Macromolecular parameters of photo layers of
gelatine laser-cross-bonded with salts of acrylic composition

Cross-bonding polymerizing component	Parameters of cross-bonded composition				
	$p, g \cdot cm^{-3}$	Q_{max}	K_{swell}, S^{-1}	$dQ/d, S^{-1}$	M^C, mol
Ca(CH$_2$=CH-C	1.71	2.73	$1.30 \cdot 10^{-5}$	$2.60 \cdot 10^{-4}$	11700
Cu(CH$_2$=CH-C	1.89	2.52	$4.60 \cdot 10^{-5}$	$1.10 \cdot 10^{-4}$	13600

Sensitometric studies of photo layers based on acrylates of metals show (Table 25) that their optical characteristics are close to or even better than those of the photolayer of gelatine with MBAD.

Table 25. Optical characteristics of photo layers of
gelatine laser-cross-bonded with salts of acrylic acid

Cross-bonding polymerizing component	Parameters of cross-bonded composition h, %	$E, mJ \cdot cm^{-2}$	$S, m^{-2} \cdot J$	$1/d, mm^{-1}$
Ca(CH$_2$=CH-C	33	50	$1.20 \cdot 10^{-3}$	1580
Cu(CH$_2$=CH-C	33	28	$2.10 \cdot 10^{-3}$	1210

Furthermore, due to the fact that the macrocomplex of copper (II) also possesses the chromophormic properties with relation to the radiation of He-Ne laser achievement of the standard D^2 33% requires a minimum exposition of only 28 mJ/cm^2.

The general scheme of the physical and chemical processes occurring in these composite photo layers are presented in Figure 19.

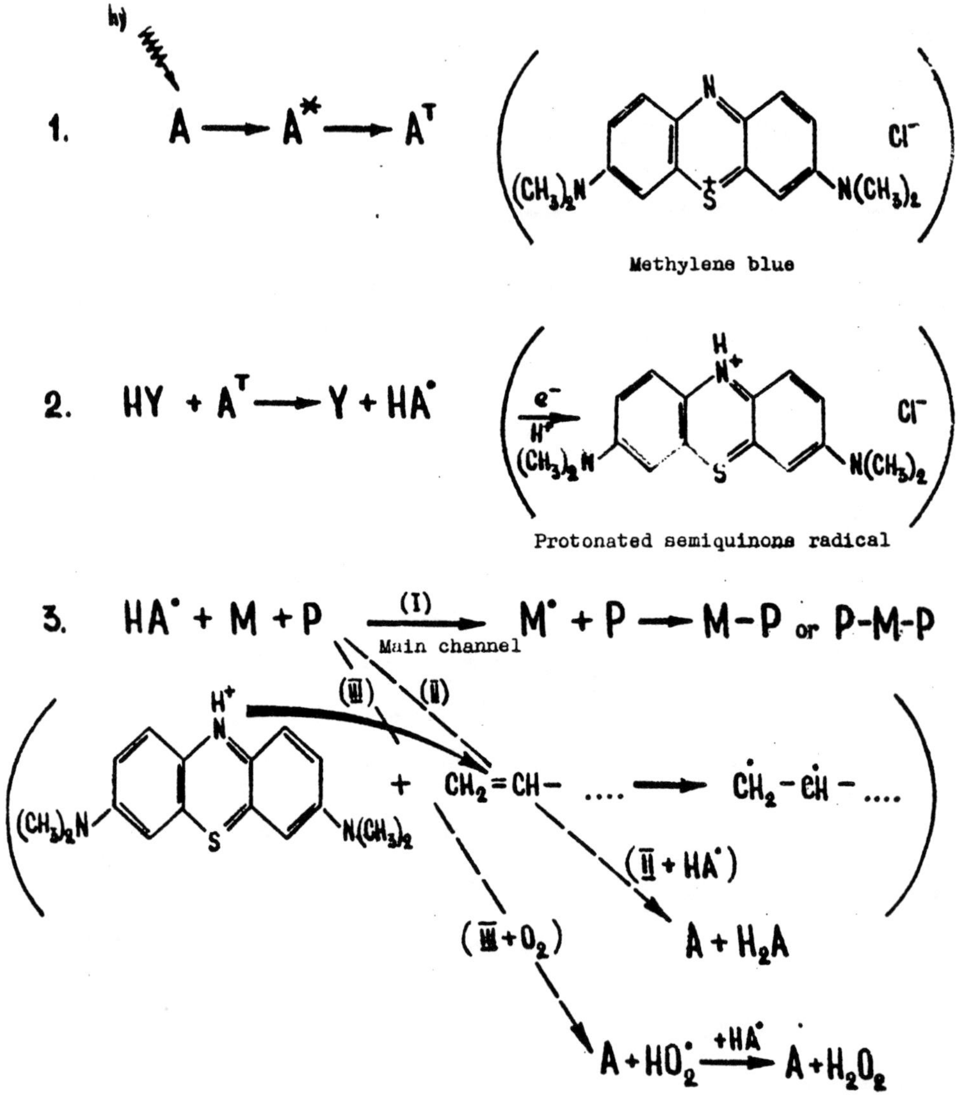

Figure 19. General scheme of processes of laser-assisted cross-linking of acrylic acid compounds with gelatine with participation of sensibilizer and donor of electron.

Having absorbed the quantum of the laser radiation, the molecule of the sensibilizer, (methylene blue -A), goes over to the excited state - A^X and then in the triplet state A^T. The latter interacts with the electron donor, triethanolamine My and forms a semichion taking part simultaneously in several reactions (I+II+III).

But the main channel (I) is the interaction with the cross-bonding component, monomeric compound of acrylic acid (M) with the formation of radical (M) due to opening of the p-bond in it, following this the radical M interacts with gelatine (p-polymer) with the formation of the introduced or cross-linked structure (M=P).

REFERENCES

1. Volkov, S.V., *J. Inorg. Chem.*, 25 No. 1, p. 87-96 (1980).
2. Volkov, S.V., *Theor. and Experiment. Chem.*, 8 No. 1, p. 3-7 (1982).
3. Volkov, S.V. *Ukr. Chem. Journ.*, 48 No. 7, p. 675-681 (1982).
4. Volkov, S.V., "High Temperature Coordination Chemistry," Proc. 29 IUPAC Congress, FRG, Koln, p. 60 (1983).
5. Volkov, S.V. *J. Inorg. Chem.*, 29 No. 2, p. 407-415 (1984).
6. Volkov, S.V., *J. Inorg. Chem.*, 31 No. 11, p. 2748-2757 (1986).
7. Yatsimirsky, K.B., *J. Inorg. Chem.*, 12 No. 12, p. 3226-3227 (1967).
8. Volkov, S.V., Grishchenko, N.F., Delimansy, Yu.K. "Coordination chemistry of molten salts," Kiev, Naukova Dumka, 1977, p. 332.
9. S.V. Volkov, V.A. Zasukha, "Quantum chemistry of coordination condensed systems," Kiev, Naukova Dumka, 1985, p. 296.
10. Smirnov, M.V., "Electrode potentials in molten chlorides," M. *Nauka*, 1973, p. 247.
11. Volkov, S.V. *Pure and Appl. Chem.*, 59 No. 9, p. 1151-1164 (1987).
12. Bockris, J.O'M., Inman, D., Reddy, A., Srinivasan, S.G., *J. Electroanalyt. Chem.*, 5, N6, p. 476 (1983).
13. "Proc. I Internat. Sympos. Molten Salt Chem. and Technology," Ed. N. Watanabe, Japan, Kyoto, 1983, p. 513.
14. K.B. Yatsimirsky, Vasilyev, "Constants of instabilities of complex compounds," M. Russia AS Publishers, 1959, p. 206.
15. S.V. Volkov, K.B. Yatsimirsky, "Spectroscopy of molten salts," Kiev, Naukova Dumka, 1977, p. 225.
16. Clarke, J.H., Hartley, P.J., Kuroda, Y., *J. Phys. Chem.*, 76, p. 1831 (1972).
17. Liehr, A.D., Ballhausen, C.J. *Ann. Phys. (N.-Y.)*, 6, p. 174 (1959).
18. Mirny, V.N., Prisyazhny, V.D., Mirnaya, T.A. in Col. "Thermodynamic and electrochemical properties of ionic melts," Kiev, Naukova Dumka, 1984, p. 54.
19. Kravtsov, V.I., "Equilibrium and Kinetics of electrode reactions," St. Petersburg, Chimiya, 1985, p. 247.
20. Shapoval, V.I., "Kushkhov, Kh.B., Electrocatalysis and electrocatalytic processes," Kiev, Naukova Dumka, 1986, p. 152.

21. Kirillov, S.A., Suynko, L.L., "Proceedings of the IX-th All-Union Conference on physical chemistry and electrochemistry of ionic melts and solid electrolytes," Sverdlovsk, 1987, vol. I, p. 230.
22. Volkov, S.V., *Coordination Chemistry*, Vol. 15, No. 6, p. 723, 1989.
23. Delimarsky, Yu.K., Markov, B.F., "Electrochemistry of molten salts," M. Metallurgizdat, 1960, p. 325.
24. Dye, H.A., Gruen, D.M., *Inorg. Chem.*, 3, No. 6, p. 836 (1964).
25. Emmenegger, F.P., Rengier, R., "Abstr. VIII Internat. Symp. on Solute-Solute-Solvent Interactions," FRG, Regensburg, p. 88, 1987.
26. Volkov, S.V., Fokina, Z.A., Timoshchenko, N.I. *Rev. de Chimie Min.*, 20, p. 776 (1983).
27. Fokina, Z.A., Synthesis, "Structure and Application of Halogencharcogenide Complex Compounds of Noble Metals," Kiev, IONCh As Ukrainian, M. Sc. thesis, 1987, p. 52.
28. Fokina, Z.A., Timoshchenko, N.I., Volkov, S.V., *Coord. chemistry*, 5, No. 3, p. 443, 1979.
29. Vokov, S.V., Timoshchenko, N.I., Fokina, Z.A., Buryak, N.I. *Ukr. chem. journ.*, 46, No. 8, p. 563, 1977.
30. Vokov, S.V., Fokina, Z.A., Pekynyo, V.I. *Timishchenko N.I.*, *Ukr. Chem. Journ.*, 46, No. 8, p. 213, 1977.
31. Vokov, S.V., Pekynyo, V.I., Fokina, Z.A., *Ukr. chem. journ.*, 47, No. 10, p. 1020, 1981.
32. Fokina, Z.A., Volkov, S.V., Baranovsky, I.B., Pekynoy, V.I., *J. Inorg. chem.*, 26 No. 6, p. 1835, 1981.
33. Fokina, Z.A., Pekhnyo, V.I., Timoshchenko, N.I., Volkov, S.V., *Ukr. chem. journl*, 47, No. 9, p. 1215, 1981.
34. Fokina, Z.A., Lapko, V.F., Volkov, S.V. et al. *Ukr. chem. journ.*, 51, No. 6, p. 573, 1985.
35. Fokina, Z.A., Timoshchenko, N.I., Lapko, V.F., Volkov, S.V., *Ukr. chem. journl.*, 48, No. 10, p. 1014, 1982.
36. Grafov, A.V., Synthesis, "Structure and Application of Complexes of Ruthenium, Rhodium and Palladium Chlorides with Chalcogen Chlorides," D. Sc. (Chem.) thesis, Kiev, IONKh AS Ukrainian, 1988.
37. Volkov, S.V., Mazurenko, E.A. "Betadiketonate Complexes of Metals in Gas Phase and Gas-Transport Reactions with their Participation," XII-th Mendeleevian Congress on general and applied chemistry, Col. Inorganic Chemistry, Baku, 1981, p. 84.
38. Mazurenko, E.A. "Synthesis, Structure and Properties of High-Volatile B-Diketonates of Metals and Gas-, Heterophase Processes with their Participation," M. Sc. (chem) thesis, Kiev, IONKh AS Russian, 1987, p. 48.
39. Volkov, S.V., Mazurenko, E.A., Zheleznova, L.I., *Ukr. chem. journ.*, 1978, Vol. 44, No. 6, p. 289-291.
40. Volkov, S.V., Mazurenko, E.A., Bublik, Zh.N., *Ukr. chem. journ.*, 1978, Vol. 44, No. 6, p. 570-573.
41. Mazurenko, E.A., Gerasimchuk, A.I., Volkov, S.V., *Theor. and exp. chem.*, 1978, vol. 15, No. 2, p. 220-226.

42. Bublik, Zh.N., Mazurenko, E.A., Volkov, S.V., *Ukr. Chem. Journ.*, 1978, vol. 44, No. 11, p. 1214-1217.
43. Mazurenko, E.A., Bublik, Zh.N., Volkov, S.V. *Ukr. chem. journ.*, 1979, Vol. 45, No. 7, p. 591-596.
44. Gerasimchuk, A.I., Volkov, S.V., Mazurenko, E.A., Maslov, V.G. *Coordination chemistry*, 1979, Vol. 5, No. 3, p. 360-366.
45. Volkov, S.V., Zheleznova, L.I., Mazurenko, E.A., *Coord. chem.*, 1979, Vol. 5, No. 3, p. 412-416.
46. Volkov, S.V., Mazurenko, E.A., Zheleznova, L.I., *Coord. chem.*, 1980, vol. 6, No. 1, p. 86-88.
47. Volkov, S.V., Gerasimchuk, A.I., Mazurenko, E.A., *Journ. Struct. Chem.*, 1980, Vol. 21, No. 2, p. 168-171.
48. Volkov, S.V., Larin, G.M., Mazurenko, E.A., Zub. V. Ya., *Coord. Chem.*, 1980, Vol. 6, No. 3, p. 344-347.
49. Gerasimchuk, A.I., Mazurenko, E.A., Volkov, S.V., *Theor. and exp. chem.*, 1981, Vol. 17, No. 2, p. 249-253.
50. Gerasimchuk, A.I., Mazurenko, E.A., Volkov, S.V., *Ukr. chem. journ.*, 1982, Vol. 48, No. 8, p. 787-790.
51. Volkov, S.V., Zub, V.Ya., Larin, G.M., Mazurenko, F.A., *Coord. chem.*, 1982, Vol. 8, No. 9, p. 1333-1336.
52. Volkov, S.V., Zub, V.Ya., Larin, G.M., Mazurenko, F.A., *Coord. chem.*, 1983, Vol. 9, No. 1, p. 26-30.
53. Mazurenko, E.A., Miropolskaya, L.E., Volkov, S.V., *Ukr. chem. journ.*, 1983, Vol. 49, No. 7, p. 682-687.
54. Bogdanov, V.A., Volkov, S.V., Gerasimchuk, A.I., Mazurenko, E.A., *Coord. chem.*, 1984, Vol. 10, p. 1346-1352.
55. Bublik, Zh.N., Volkov, S.V., Mazurenkko, E.A., *J. inorg. chem.*, Vol. 29, No. 1, 1984, p. 132-137.
56. Volkov, S.V., Zheleznova, L.I., Mazurenko, E.A., *Coord. chem.*, 1983, Vol. 9, No. 10, p. 1338-1341.
57. Volkov, S.V., Vorobyev, A.V., Gerasimchuk, A.I., Mazurenko, E.A., *Ukr. chem. journ.* 1983, Vol. 49, No. 9, p. 899-901.
58. Volkov, S.V., Gurko, A.F., Lutoshkin, V.I., Mulenko, S.A., Botsman, A.V., *Visnik Ukr.*, AS 50, No. 8, p. 65-77, 1986.
59. Volkov, S.V., Gurko, A.F., Lutoshkin, V.I., *Theor. and exp. chem.* 17, No. 4, p. 564-567, 1981.
60. Volkov, S.V., Gurko, A.F., Lutoshkin, V.I., *Ukr. chem. journ.*, 48, No. 5, p. 451-453, 1982.
61. Gurko, A.F., Lutoshkin, V.I., Volkov, S.V., *Ukr. chem journ.*, No. 4, p. 376-382, 1983.
62. Lutoshkin, V.I., Marchenko, L.V., Volkov, S.V. et al., *Ukr. chem. journ.*, 49, No. 5, p. 485-488.
63. "Application of lasers in spectroscopy and photochemistry," Edited by K. Mur. - M. Mir, 1983, p. 45.
64. Bauer, S.H., Chick, K.R., *Chem. Phys. Lett.*, 45, No. 3, p. 529-532 (1977).
65. Jakobs, R.R., Krupke, W.F., Hessler, J.P., Carnall, W.T., *Opt. Comuns.*, 21, No. 3, p. 395-398 (1977).

66. Jakobs, R.R., Krupke, W.F., *Appl. Phys. Lett.*, 34, No. 8, p. 497-499 (1979).
67. Papatheodorou, L.N., Berg, R.W., *Chem. Phys. Lett.*, 75, No. 3, p. 483-487 (1980).
68. Mulenko, S.A., Lutoshkin, V.I., Kostromina, N.A., Volkov, S.V., *Theor. and exp. chem.*, 21, p. 367-370, 1985.
69. Lutoshkin, V.I., Popov, O.F., Volkov, S.V., *Theor. and exp. chem.* 17, No. 6, p. 696-699, 1981.
70. Lutoshkin, V.I., Volkov, S.V., Botsman, A.V., *Ukr. chem. journl.* 50, No. 3, p 250-254, 1984.
71. Volkov, S.V., Botsman, A.V., Shevchenko, A.P., *Ukr. chem., journ.*, 53, No. 5, p. 494-497, 1987.
72. Volkov, S.V., Botsman, A.V., Lutoshkin, V.I., Pomagaylo, A.D., *Ukr. chem. journ.*, 54, No. 9, p. 899-903, 1988.
73. Botsman, A.V., Volkov, S.V., Lutoshkin, V.I., Pomogaylo, A.D., *Ukr. chem. journ.*, 55, No. 4, p. 341-345, 1989.

Subject Index

A

Acyl Ferrocene Derivatives, Metallochelates Based on Thisemicarbazones and Hydrazones of 85-88

Ambivalent Ligands, Isomerism in Simple Complexes of Metal Cations with 1-74

Anhydrous High-Boiling Solvents, Chemistry of Coordination Compounds in, 183

Anions NO2-, NS2-, PO2-, and PS2-, Molecules O3SO2, S2O, S3 and Their Analogs,29

B

B-Diketonates of Metals, Volatility of, 210-217

Boron Hydride Structures, Formation of Polyhedral, 109-110

Boron in Coordination Chemistry, 107-150

C

Chalkogenides YAY'(Y,Y'=0 and S,A=B, Al, C and Si, etc.), 24-26

Charge of Cation and its Chemical Environment,17-20

Chelate Formation, Structural Modification of Ferrocene by the Method of, 91-98

Chemical Bonds, Structure and Peculiar Features of, 111-114

Chemical Properties and Their Relationships With Structure, 138

Chemical Studies of Systems and Synthesis of Complexes, 135

Chemistry of Coordination Compounds in Anhydrous High-Boiling Solvents, 183

Chemistry of Coordination Compounds in the Vapor (Gas) Phase, 197

Closo-Dodecaborate-Halides, Complex, 134-139

Cluster Boron Compounds, 108-124

Cluster Compounds of Boron in Coordination Chemistry, 107-150

Co(II), Ni(II), Cu(II) and Ln(III), Complex Compounds of Mn(II), Fe(II), 128-131

Complex Anion CuB10H10, 144

Complex Closo-Dodecaborate-Halides, 134-139

Complex Compounds of Mn(II), Fe(II), Co(II), Ni(II), Cu(II) and Ln(III), 128-131

Complex Compounds of Palladium (II) and Platinum (II), 126-127

Complex Compounds of Palladium (II) and Platinum (II), 126-127

Complex Compounds of Pb(II), 143

Complex Compounds of Pt(II), Pd(II) and Cu(I), 141

Complex Compounds of Transition Metals With Outer-Sphere Polyhedral Anions,125-133

Complex Compounds of Uranyl UO2+ and U(IV), 132-133

Complex Formation, Complex-Cluster Model of Structure of Molten-Salt Systems with, 158-162

Complex-Cluster Model of Structure of Molten-Salt Systems with Complex Formation, 158-162

Complexes LiAB with 14 Valent Electrons, 22-23

Complexes M[ABC] of Triatomic Ligands ABC with 16 Valent Electrons, 24

Complexes M[AB] with Diatomic AB Ligands Containing 10 Valent Electrons, 7-11

Complexes of Triatomic ABC Ligands With 18 and 20 Valent Electrons, 29

Computational Approximation and Quality of Nonempirical Calculations, 4-6

Coordination Chemistry, Cluster Compounds of Boron in, 107-150

Coordination Chemistry, High Temperature, 151-233

Coordination Compounds in Anhydrous High-Boiling Solvents, Chemistry of, 183

Coordination Compounds in Laser Chemistry [58-72], 221

Coordination Compounds in Molten Salts, 167-178

Coordination Compounds in Molten Salts,Laser Chemistry of, 224

Coordination Compounds in Polymerizing and Solid Phases, 227-232

Coordination Compounds in the Vapor (Gas) Phase, Chemistry of, 197

Crystal-Chemistry Stability Criterion, Structure and, 136

Cu(I), Complex Compounds of Pt(II), Pd(II) and, 141

Cu(II) and Ln(III), Complex Compounds of Mn(II), Fe(II), Co(II), Ni(II), 128-131

CuB10H10, Complex Anion, 144

Cyanates and Thiocyanates, 27-28

D

Di- and Triatomic Ligands AB and ABC, 7-11

Diatomic AB Ligands Containing 10 Valent Electrons, 7-11

Diatomic Ligands with 12 Valent Electrons, 20-21

Dimers (MAB)2, 36-38

Dinuclear and other Complexes, Formation of Heteronuclear, 189-195

Discrete Complex Ion in Melts as a Quadrinuclear Heteronuclear System, 163-166

C

Fe(II), Co(II), Ni(II), Cu(II) and Ln(III), Complex Compounds of Mn(II), 128-131

Ferrocene by the Method of Chelate Formation, Structural Modification of, 91-98

Ferrocene-Based Metallochelates, 75-106

Ferrocene-Diketones, Transition Metal Complexes With , 76-84

Formation of Heteronuclear, Dinuclear and other Complexes, 189-195

Formation of Polyhedral Boron
    Hydride Structures, 109-110
Fragments LiAB of Diatomic
    Ligands with 12 Valent
    Electrons, 20-21

G
Gas Phase, Laser Chemistry of
    High Coordination
    Compounds in, 222
Gas-Transport, Homophase,
    Heterophase Processes With
    Participation of High-Volatile
    B-Diketonate Complexes [38],
    218-220

H
HAB->ABH, PES of
    Rearrangements MAB->ABM
    (M=Li and Cu) and,12-16
Hetermetallic Ferrocene-Containing
    Metallochelates, 89-90
Heteronuclear, Dinuclear and other
    Complexes, Formation of, 189-
    195
Heterophase Processes With
    Participation of High-Volatile
    B-Diketonate Complexes [38],
    Gas-Transport, Homophase,
    218-220
High Coordination Compounds in
    Gas Phase, Laser Chemistry of,
    222
High Temperature Coordination
    Chemistry, 151-233
High-Volatile B-Diketonate
    Complexes [38], Gas-
    Transport, Homophase,
    Heterophase Processes With
    Participation of, 218-220
Homophase, Heterophase
    Processes With Participation
    of High-Volatile B-Diketonate

Complexes [38], Gas-
    Transport, 218-220
Hydrazones of Acyl Ferrocene
    Derivatives, 85-88

I
Ion Coordination in Molten Salts
    and Molecules in Molecular
    High-Boiling Solvents, 179-182
Ion-Molecular Complexes With Di-
    and Triatomic Ligands AB and
    ABC, 7-11
Ionic Metals, Chemistry of
    Coordination Compounds in,
    156-157
Ions M2AB+ With Two Cations,
    31-33
Ions M2ABC+, 34-35
Isomerism in Simple Complexes of
    Metal Cations with
    Ambivalent Ligands, 1-74
Isomerism of M[ABC] Salt
    Molecules and Ion-Molecular
    Complexes With Di- and
    Triatomic Ligands AB and
    ABC, 7-11

L
Laser Chemistry of Coordination
    Compounds in Molten Salts,
    224
Laser Chemistry of Coordination
    Compounds in Polymerizing
    and Solid Phases, 227-232
Laser Chemistry of High
    Coordination Compounds in
    Gas Phase, 222
Laser Chemistry [58-72],
    Coordination Compounds in,
    221
Ligands ABC With 16 Valent
    Electrons, 28

Ligands, Polyhedral Anions BnHn
    as, 140
Ln(III), Complex Compounds of
    Mn(II), Fe(II), Co(II), Ni(II),
    Cu(II) and, 128-131

M

MAB->ABM (M=Li and Cu) and
    HAB->ABH, PES of
    Rearrangements, 12-16
Melts as a Quadrinuclear
    Heteronuclear System, Discrete
    Complex Ion in, 163-166
Metal Cations with Ambivalent
    Ligands, 1-74
Metallochelates Based on
    Thisemicarbazones and
    Hydrazones of Acyl Ferrocene
    Derivatives, 85-88
Metallochelates, Ferrocene-Based,
    75-106
Metallochelates, Hetermetallic
    Ferrocene-Containing 89-90
Metals, Volatility of B-Diketonates
    of, 210-217
Mn(II), Fe(II), Co(II), Ni(II), Cu(II)
    and Ln(III), Complex
    Compounds of, 128-131
Molecular High-Boiling Solvents,
    Ion Coordination in Molten
    Salts and Molecules in, 179-
    182
Molecules in Molecular High-
    Boiling Solvents, Ion
    Coordination in Molten Salts
    and, 179-182
Molten Salts and Molecules in
    Molecular High-Boiling
    Solvents, Ion Coordination in,
    179-182
Molten Salts, Laser Chemistry of
    Coordination Compounds in,
    224

Molten Salts, Quantum-Chemistry
    and Structure and Radiative
    and Nonradiative Transitions
    of Coordination Compounds
    in, 167-178
Molten-Salt Systems with
    Complex Formation, 158-162
Monoclinic Structures and
    Transition Between Them,
    Polyhedrons B10H11- in
    Tetragonal and , 122-124
M[ABC] Salt Molecules and Ion-
    Molecular Complexes With Di-
    and Triatomic Ligands AB and
    ABC, 7-11
Ni(II), Cu(II) and Ln(III), Complex
    Compounds of Mn(II), Fe(II),
    Co(II), 128-131
Nonempirical Calculations,
    Computational Approximation
    and Quality of, 4-6
Nonradiative Transitions of
    Coordination Compounds in
    Molten Salts, Quantum-
    Chemistry and Structure and
    Radiative and 167-178

O

Outer-Sphere Polyhedral Anions,
    Complex Compounds of
    Transition Metals With, 125-
    133

P

Palladium (II) and Platinum (II),
    Complex Compounds of, 126-
    127
Pb(II), Complex Compounds of,
    143
Pd(II) and Cu(I), Complex
    Compounds of Pt(II), 141

PES of Rearrangements MAB->ABM (M=Li and Cu) and HAB->ABH, 12-16

Physical and Chemical Properties and Their Relationships With Structure, 138

Physical and Chemical Studies of Systems and Synthesis of Complexes, 135

Platinum (II), Complex Compounds of Palladium (II) and, 126-127

Polyhedral Anions BnHn as Ligands, 140

Polyhedral Boron Hydride Structures, 109-110

Polyhedrons B10H11- in Tetragonal and Monoclinic Structures and Transition Between Them, 122-124

Polymerizing and Solid Phases, Laser Chemistry of Coordination Compounds in, 227-232

Pt(II), Pd(II) and Cu(I), Complex Compounds of, 141

Q

Quadrinuclear Heteronuclear System, Discrete Complex Ion in Melts as a, 163-166

Quantum Chemistry, Spectroscopy and Structure of Volatile B-Diketonate Metals and Vapor(Gas) Phase [38,41,44,47,49,50,54], 200-209

Quantum-Chemistry and Structure and Radiative and Nonradiative Transitions of Coordination Compounds in Molten Salts, 167-178

R

Radiative and Nonradiative Transitions of Coordination Compounds in Molten Salts, 167-178

Reaction Capacity, 115-116

S

Simple Complexes of Metal Cations with Ambivalent Ligands, 1-74

Simultaneous Exhibition of Properties of Solvent-Ligand-Reagent, 183-188

Solid Phases, Laser Chemistry of Coordination Compounds in Polymerizing and, 227-232

Solvato-Metallurgical Aspects, 196

Solvent-Ligand-Reagent, Simultaneous Exhibition of Properties of, 183-188

Solvents, Chemistry of Coordination Compounds in Anhydrous High-Boiling, 183

Solvents, Ion Coordination in Molten Salts and Molecules in Molecular High-Boiling, 179-182

Spectroscopy and Structure of Volatile B-Diketonate Metals and Vapor(Gas) Phase [38,41,44,47,49,50,54], 200-209

Structural Modification of Ferrocene by the Method of Chelate Formation, 91-98

Structure and Crystal-Chemistry Stability Criterion, 136

Structure and Peculiar Features of Chemical Bonds, 111-114

Structure and Radiative and Nonradiative Transitions of

Coordination Compounds in
Molten Salts, 167-178
Superelectron deficient Structures,
117-121
Synthesis of Complexes, Physical
and Chemical Studies of
Systems and, 135

T
Tetragonal and Monoclinic
Structures and Transition
Between Them, 122-124
Thiocyanates, Cyanates and 27-
28
Thisemicarbazones and
Hydrazones of Acyl Ferrocene
Derivatives, 85-88
Transition Metal Complexes With
Ferrocene-Diketones,76-84
Transition Metals With Outer-
Sphere Polyhedral Anions,125-
133
Triatomic ABC Ligands With 18
and 20 Valent Electrons, 29
Triatomic Ligands AB and ABC,
7-11

Triatomic Ligands ABC with 16
Valent Electrons, Complexes
M[ABC] of, 24

U
Uranyl UO2+ and U(IV), Complex
Compounds of , 132-133

V
Vapor (Gas) Phase, Chemistry of
Coordination Compounds in
the, 197
Vapor(Gas) Phase
[38,41,44,47,49,50,54],
Quantum Chemistry,
Spectroscopy and Structure of
Volatile B-Diketonate Metals
and, 200-209
Volatile B-Diketonate Metals and
Vapor(Gas) Phase
[38,41,44,47,49,50,54], 200-
209
Volatility of B-Diketonates of
Metals, 210-217